T0275887

Conceptual Breakthroughs in Ethology and Animal Behavior

Conceptual Breakthroughs in Ethology and Animal Behavior

Michael D. Breed

University of Colorado, Boulder, CO, United States

Academic Press is an imprint of Elsevier
125 London Wall, London EC2Y 5AS, United Kingdom
525 B Street, Suite 1800, San Diego, CA 92101-4495, United States
50 Hampshire Street, 5th Floor, Cambridge, MA 02139, United States
The Boulevard, Langford Lane, Kidlington, Oxford OX5 1GB, United Kingdom

British Library Cataloguing-in-Publication Data
A catalogue record for this book is available from the British Library

Library of Congress Cataloging-in-Publication Data
A catalog record for this book is available from the Library of Congress

ISBN: 978-0-12-809265-1

For Information on all Academic Press publications
visit our website at https://www.elsevier.com/books-and-journals

Publisher: Sara Tenney
Acquisition Editor: Kristi Gomez
Editorial Project Manager: Pat Gonzalez
Production Project Manager: Karen East and Kirsty Halterman
Designer: Matthew Limbert

Typeset by MPS Limited, Chennai, India

Dedication

For my wife Cheryl, for her love and support

Contents

Preface

This book attempts to capture a snapshot of the development of animal behavior as a field of scientific inquiry, using key moments in the field to illustrate its chain of development. Examination of the thread of animal behavior science is a fascinating platform for understanding how science and society interact. Animal behavior has been replete with scientists who wished to project their findings and theories into public debates on topics such as social welfare, violence, aggression and criminality, and women as caretakers/nurturers. The ambition of some scientists to be more than toilers in the vineyards of knowledge has helped to illuminate societal conversations but it has also, on occasion, brought ire, criticism, and scorn onto those who put themselves into the public eye. Unfortunately, sometimes, the vitriol aimed at scientists engaged in this discourse has splashed over onto those who prefer to explore the wonders of nature while leaving politics and philosophy to others. The sociobiology controversy, specifically, caused many scientists to rearticulate what they called themselves—behavioral ecologists rather than sociobiologists—to avoid being embroiled and to evade interjection into the seemingly endless nature–nurture controversy. In animal behavior, scientists who have pushed far beyond their own findings into the public arena have exacted a toll on their less adventurous colleagues, who unfailingly are asked to explain and justify others' statements that extrapolate from observations of animals, be they ants, bees, dolphins, or monkeys, to human behavior, culture, and society. From the perspective of the 2010s, I hope that this history makes for entertaining reading. Certainly not all of the findings that I highlight have created such heat, but the spinning of the thread and weaving of the fabric of the scientific world of animal behavior from its earlier origins to its present sophistication is fascinating, and I have tried to do justice to this intriguing story.

I am indebted to John Avise's 2014 book on evolutionary genetics, which served as both an inspiration for this project and as a model for its execution. I am also grateful to the hundreds of students who have taken the course in animal behavior at the University of Colorado, Boulder, from me. Teaching is a great laboratory in which, through trial and error, important can be sifted from mundane, fascinating from stultifying, and impactful from dull. I am also highly appreciative of the graduate students who have been part of the teaching team in the course over the years, as they have served as critics, muses, sounding boards, reality checkers, and fact checkers.

Michael D. Breed
Boulder, Colorado, September 2016

REFERENCE

Avise, J.C., 2014. Conceptual Breakthroughs in Evolutionary Genetics: A Brief History of Shifting Paradigms. Elsevier, New York, p. 164.

Introduction

Animal behavior is one of the most dynamic and exciting areas in science. Its roots grew in ethology and comparative psychology, but physiology, ecology, and evolution inform many of the major concepts of animal behavior.[1] In fact, animal behavior has entranced humans since well before any written record of our activities. My aim in writing this book was to integrate the history of the study of animal behavior by finding key moments that have shaped our understanding and application of animal behavior in our lives.

This book was inspired by John Avise's 2014 book, *Conceptual Breakthroughs in Evolutionary Genetics*. I am grateful to have been able to use his work as a model. The concept is to select the key points in the field's history—pivot points at which a new insight impels future work. Each key point is highlighted with an explanation of the concept and an evaluation of how important it was in shaping the field. I have followed Avise's lead in giving each concept or breakthrough an impact score on a scale of 1–10, but the scores should be taken with a grain of salt, because all of the chapters in this book represent important moments in the study of animal behavior.

Human interest in animal behavior reflects the very strong survival value for us gained from understanding the behavior of our predators, our parasites, our prey, and perhaps most of all, the behavior of other humans. The study of animal behavior need not be anthropomorphic. In fact, imbuing animals with human traits is a major pitfall in the study of behavior; but we also need to not consider human capabilities as unreachable by other species. In line with this, we must always recognize that humans exist as just another animal, not separated or somehow segregated from the evolutionary processes that shape any animal species. So while animals should not be viewed as furry or feathery humans, we should also not under-credit the abilities of other species by regarding them as lower forms of life. Human abilities in behavior and sensory perception evolved in a continuum from what other animals could do and perceive. Any one animal species is not above or below another species in an evolutionary order; rather, all species are the result of the cumulative effects of the evolutionary process.

1. Stated in this way, this sounds much like the grand synthesis claimed by E. O. Wilson (1975) for sociobiology in his book, *Sociobiology*. As the present book shows, the synthesis evolved over the last 150 years, with students and scholars of animal behavior never being unaware of the importance of understanding physiology, evolution, and ecology in order to interpret how and why animals behave.

The main task in writing this book was to select the appropriate concepts and historical moments to highlight. One source was scientific journals in the discipline. Scientists working on ethology and animal behavior founded journals focused on these topics, and these journals facilitated development of the fields. Karl von Frisch, discoverer of the dance language of bees, cofounded the *Zeitscrift fur Vergleichende Physiologie* (now called *Journal of Comparative Physiology*) in 1924. The *Zeitschrift fur Tierpsycholgie* (*Journal of Animal Psychology*, now called *Ethology*) was founded in 1937 and attracted work by leading ethologists. *Animal Behaviour*, copublished by the British and American scientific societies for studying animal behavior, was founded as the *British Journal of Animal Behaviour* in 1953. *Behaviour* was founded by Niko Tinbergen and W. H. Thorpe in 1948. Founding a journal was a function of the leadership of these men in the field and these journals were the leading outlets for ethological studies until the 1970s.

Behavioral ecology came into its own as a field in the 1970s, fueled by the desire of some scientists who studied animal behavior to focus on ultimate evolutionary causes of behavior, rather than on proximate functional regulation of behavior. This development stimulated the founding of the journal *Behavioral Ecology and Sociobiology* with Edward O. Wilson and Bert Holldobler as coeditors. Later the journal *Behavioral Ecology*, founded in 1990, also began serving this niche.

From the 1970s until recently, *Animal Behaviour* was the leading journal for proximate studies of behavior and *Behavioral Ecology and Sociobiology* and *Behavioral Ecology* were the leading journals for studies of the ultimate causes of behavior. The journal landscape has shifted substantially with the advent of online publication and open access journals, but for the present these three journals remain very important places for publication of behavioral studies.

Finding concepts and historical moments is not as easy as it sounds, and certainly a large number of pitfalls exist in its application. The most difficult task is to assign a date and specific study to a concept. Generally speaking, science builds on itself; I think of each addition to scientific knowledge as another brick in the wall of understanding. But some bricks are larger and perhaps especially prominent in the architecture of the wall—these might be thought of as keystones. If a behavior was first noted by a 19th century naturalist, built into a concept by an ethologist in the 1940s or 1950s, elaborated into a theory complete with mathematical models in the 1970s, and then reworked with new insights in the 1990s, what is the key moment for that concept? Really, none of these is the single crucial element as each depends on the others for its existence. The structure of this book depends on choosing fundamental moments, so I have relied on my subjective assessments of the impacts of individual contributions and tools like citation rates to select important moments, or pivot points, in the study of animal behavior.

Another problem is ascertaining the lasting value of work. Perceptions of the importance of work in the years after its initial publication may be distorted by fads and the relentless self-promotion employed by some investigators. Sometimes research directions do not ultimately pan out even if they receive considerable initial attention. These effects cause the luster of some work to fade over time. Because of the unknowns about the impacts of more current research, I have highlighted relatively few moments since 2000. We simply do not yet know the impact of much of the work published in the last decade and a half. It is preposterous to think that citations of published work over a year, 2 years, or 3 years postpublication say anything about the enduring impact of scientific discovery. If someone writes a book with the same approach as I have taken here in the year 2040, likely the ideas developed between 2010 and 2030 will form the centerpieces of the field, as by then perceptions of the enduring nature of certain contributions will have been formed.

In addition to my own judgments of the impact of concepts and specific publications, I consulted a variety of online sources, including *Web of Science*, *Google Scholar*, and *Research Gate*. There are also a number of fine resources in print or online that analyze the history of thought in ethology, animal behavior, and comparative psychology (Dewsbury, 1984, 1989; Drickamer and Dewsbury, 2009). These also include the book, *Foundations of Animal Behavior: Classic Papers with Commentaries* (Houck and Drickamer, 1996), a set of essays published in 2013 in the journal *Animal Behaviour* that honored the 60th anniversary of the journal, a history of ethology by Egerton (2016), and Klopfer and Hailman's textbook (1972), which takes a historical approach.

Choosing older concepts and specific works for inclusion was relatively easy as time has tested the value and sustainability of the concepts and ideas. While very old work is likely not cited in contemporary scientific publications, its remanence endures and it does not take much sleuthing to realize that the roots of current science are in books and journal articles from the 19th century and the first half of the 20th century. Certainly I have missed concepts or findings that others might see as key early developments in the field, but my selections reflect consideration of the choices of older works by Drickamer and Houck (1996) and the history of the field as outlined by Klopfer and Hailman (1972).

When I started the outline for this book, I called on the ability in *Web of Science* to sort the articles published on a specific topic or in a particular journal by the number of times each article had been cited. This yielded two lists, one of the most cited articles on the topic of animal behavior and the other of the most cited articles published in *Animal Behaviour* since its inception (Tables 1 and 2). I also checked citation lists for *Behavioral Ecology and Sociobiology* and *Behavioral Ecology*. These lists then informed selections of concepts for chapters in this book.

TABLE 1 Papers With At Least 2000 Citations in a *Web of Science* Search Using the Key Terms *Animal* and *Behavior/Behaviour,* Edited for Relevance of Results, November 2015

1. Genetics of *Caenorhabditis elegans*. 1974. Brenner, S. Genetics 77: 71–94 *7435 citations*
2. Observational study of behavior—sampling method. 1974. Altmann, J. Behaviour 49: 227–267 Published: *7178 citations*
3. The genome of the social amoeba *Dictyostelium discoideum*. 2005. Eichinger, L; Pachebat, JA; Glockner, G; et al. Nature 435: 43–57 *4749 citations*
4. Behavioral decisions made under the risk of predation—a review and prospectus. 1990. Lima, SL; Dill, LM. Canadian Journal of Zoology-Revue Canadienne De Zoologie 68: 619–640 *4053 citations*
5. [a]Place navigation impaired in rats with hippocampal lesions. 1982. Morris, R G M; Garrud, P; Rawlins, J N P; et al. Nature 297: 681–683 Published: *3850 citations*
6. Evolution of reciprocal altruism. 1971. Trivers, RL. Quarterly Review of Biology 46: 35–57 *3758 citations*
7. Emotion circuits in the brain. 2000. LeDoux, JE. Annual Review of Neuroscience 23: 155–184 *3677 citations*
8. Ecology, sexual selection, and evolution of mating systems. 1977. Emlen, ST; Oring, LW. Science 197: 215–223 *3579 citations*
9. [a]Developments of a water-maze procedure for studying spatial-learning in the rat. 1984. Morris, R. Journal of Neuroscience Methods 11: 47–60 *3466 citations*
10. Community structure in social and biological networks. 2002. Girvan, M; Newman, MEJ. Proceedings of the National Academy of Sciences of the United States of America 99: 7821–7826 *3203 citations*
11. Validation of open:closed arm entries in an elevated plus-maze as a measure of anxiety in the rat. 1985. Pellow, S; Chopin, P; File, SE; et al. Journal of Neuroscience Methods 14:149–167 *3100 citations*
12. The evolution of cooperation. 1981. Axelrod, R; Hamilton, WD. Science 211: 1390–1396 *2908 citations*
13. On territorial behavior and other factors influencing habitat distribution in birds. Part 1 theoretical development. 1969. Fretwell SD; Lucas HL Jr. Acta Biotheoretica 19: 16–36 *2525 citations*
14. Mate selection—selection for a handicap. 1975. Zahavi, A. Journal of Theoretical Biology 53: 205–214 *2431 citations*
15. Optimal foraging, marginal value theorem. 1976. Charnov, EL. Theoretical Population Biology 9: 129–136 *2406 citations*
16. Organisms as ecosystem engineers. 1994. Jones, CG; Lawton, JH; Shachak, M. Oikos 69: 373–386 *2376 citations*
17. Epigenetic programming by maternal behavior. 2004. Weaver, ICG; Cervoni, N; Champagne, FA; et al. Nature Neuroscience 7: 847–854 *2370 citations*

(*Continued*)

TABLE 1 (Continued)

18. Life-history tactics—review of ideas. 1976. Stearns, SC. Quarterly Review of Biology 51: 3–47 *2305 citations*

19. Mating systems, philopatry and dispersal in birds and mammals. 1980. Greenwood, PJ. Animal Behaviour 28: 1140–1162 *2235 citations*

20. Heritable true fitness and bright birds—a role for parasites. 1982. Hamilton, WD; Zuk, M. Science 218: 384–387 *2189 citations*

21. Working memory. 1992. Baddeley, A. Science 255: 556–559 *2106 citations*

22. Sperm competition and its evolutionary consequences in insects. 1970. Parker, GA. Biological Reviews of the Cambridge Philosophical Society 45: 525–567 *2034 citations*

23. Coordination of circadian timing in mammals. 2002. Reppert, SM; Weaver, DR. Nature 418: 935–941 *2006 citations*

24. Parent-offspring conflict.1974. Trivers, RL. American Zoologist 14: 249–264 *2001 citations*

[a] *These two papers are on the same topic and are often cited together.*

TABLE 2 Most Cited Papers in the Journal *Animal Behaviour*, as of July 1, 2016, *Web of Science*

Rank	Title	Authors	Volume and Pages in *Animal Behaviour*	Year Published	Number of Citations
1.	Mating systems, philopatry and dispersal in birds and mammals	Greenwood PJ	28:1140–1162	1980	2298
2.	Tests for emotionality in rats and mice—review	Archer J	21:205–235	1973	1076
3.	The logic of asymmetric contests	Smith JM, Parker GA	24:159–175	1976	1036
4.	Social behavior of anuran amphibians	Wells KD	25:666–693	1977	1032

(*Continued*)

TABLE 2 (Continued)

Rank	Title	Authors	Volume and Pages in *Animal Behaviour*	Year Published	Number of Citations
5.	The evolution of polyandry: multiple mating and female fitness in insects	Arnqvist G, Nilsson T	60:145–164	2000	908
6.	Postnatal-development of locomotion in laboratory rat	Altman J, Sudarshan K	23:896–920	1975	574
7.	Stereotypies—a critical review	Mason GJ	41:1015–1037	1991	566
8.	Competition for mates and predominant juvenile male dispersal in mammals	Dobson FS	30:1183–1192	1982	548
9.	Logical stag—adaptive aspects of fighting in red deer (*Cervus elaphus* L.)	Clutton-Brock TH, Albon SD, Gibson RM, et al.	27:211–225	1979	501
10.	Receiver psychology and the evolution of animal signals	Guilford T, Dawkins MS	42:1–14	1991	481
11.	A model for the evolution of despotic versus egalitarian societies	Vehrencamp SL	31:667–682	1983	472
12.	Cuckoos versus reed warblers—adaptations and counteradaptations	Davies NB, Brooke MD	36: 262–284	1988	460

(*Continued*)

TABLE 2 (Continued)

Rank	Title	Authors	Volume and Pages in *Animal Behaviour*	Year Published	Number of Citations
13.	The repeatability of behaviour: a meta-analysis	Bell AM, Hankison SJ, Laskowski KL	77:771–783	2009	452
14.	Ontogeny of behaviour in albino rat	Bolles RC, Woods PJ	12:427–441	1964	449
15.	Why individual vigilance declines as group size increases	Roberts G	51:1077–1086	1996	442
16.	Sexual coercion in animal societies	Clutton-Brock TH, Parker GA	49:1345–1365	1995	432
17.	Vervet monkey alarm calls—semantic communication in a free-ranging primate	Seyfarth RM, Cheney DL, Marler P	28:1070–1094	1980	429
18.	Producers and scroungers—a general-model and its application to captive flocks of house sparrows	Barnard CJ, Sibly RM	29:543–550	1981	407
19.	Territorial defence in speckled wood butterfly (*Pararge aegeria*)—resident always wins	Davies NB	26:138–147	1978	402
20.	Female mate choice in treefrogs—static and dynamic acoustic criteria	Gerhardt HC	42:615–635	1991	400

This book focuses on concepts and their impact. For the most part, the concepts trace to a particular person or group of collaborators. The chapters, with a few exceptions, are also linked to a specific scientific publication. This approach emphasizes concepts over the narratives of individual scientists, but acknowledges the importance of individuals in driving conceptual breakthroughs.

The field of animal behavior can also be understood by looking at the key individuals who contributed to the growth of scientific interest in behavior. Two books compile autobiographies of leading animal behaviorists (Dewsbury, 1989; Drickamer and Dewsbury, 2009) and are excellent sources of biographical information about key figures in the history of the field (Tables 3 and 4). The two books of autobiographies, of course, relied on individuals being alive at the times the books were prepared and being willing to write autobiographically, so some immensely important figures, like Daniel Lehrman, who died relatively young, are not represented. All of the people who wrote for these two books had tremendous impact on the study of animal behavior and on the discussions that, over the years, shaped the field. The work of some of these individuals has translated into the pivot points in animal behavior featured in this book.

TABLE 3 Autobiographies and Conceptual Contribution of Authors in Dewsbury (1989)

Scientist	Field
Baerends, Gerard P	Modern ethology, gull behavior
Dethier, Vincient G	Sensory physiology
Eibl-Eibesfeldt, Irenaus	Human ethology
Fuller, John L	Behavioral genetics
Griffin, Donald R	Bat echolocation
Hediger, Heini	Personal space
Hess, Eckhard	Imprinting and learning
Hinde, Robert A	Motivation and drive
King, John A	Critical periods for learning
Leyhausen, Paul	Ethology, behavior of fields
Lorenz, Konrad	Imprinting, public promotion of ethology
Manning, Aubrey	Ethology
Marler, Peter	Development of birdsong

(*Continued*)

TABLE 3 (Continued)

Scientist	Field
Maynard Smith, John	Game theory
Richter, Curt P	Behavioral physiology, appetite and diet
Scott, John P	Behavioral genetics, dogs
Tinbergen, Niko	Foundations of ethology
Wilson, Edward O	Social insect biology
Wynne-Edwards, Vero C	Group selection

TABLE 4 Autobiographies and Conceptual Contribution of Authors in Drickamer and Dewsbury (2009)

Alexander, Richard D	Evolution of communication, social evolution
Altmann, Jeanne	Primate behavior, research methods
Bateson, Patrick	Development
Brown, Jerram L	Avian mating systems
Clutton-Brock, Tim	Behavioral ecology
Davies, Nicholas B	Behavioral ecology
Dawkins, Marion S	Animal welfare
Dawkins, Richard	Selfish gene, public promotion of animal behavior
de Waal, Franz BM	Primate behavior
Emlen, Stephen T	Social behavior
Galef, Bennett G	Aversive learning
Gowaty, Patricia A	Evolution of sexual conflict
Hrdy, Sarah B	Primate behavior
Krebs, John R	Behavioral ecology
Orians, Gordon H	Mating systems
Parker, Geoff A	Theory
Ryan, Michael J	Signaling and mating systems
West, Meredith J	Brain, behavior and development
West-Eberhard, Mary Jane	Social evolution
Wingfield, John	Behavioral endocrinology
Zahavi, Amotz	Handicap principle

For American scientists and select non-Americans who work or worked on animal behavior and who were elected as members of the U.S. National Academy of Science (Table 5), the biographical memoirs of the Academy are very helpful resources for understanding the impacts of individuals and the concepts that they developed (http://www.nasonline.org/publications/biographical-memoirs/). Also, while there is no Nobel Prize awarded specifically for animal behavior (or for ecology), several Nobel Prize winners have been recognized for contributions to animal behavior (Table 6).

TABLE 5 Members of the United States National Academy of Sciences Whose Work At Least Partly Focuses or Focused on Animal Behavior. The NAS Is Divided Into Sections; There Is No Animal Behavior Section. The Scientists Listed Here Are Scattered Among a Number of Sections: Anthropology, Evolution, Environmental Sciences/Ecology, Systems Neuroscience, Cellular, and Molecular Neuroscience, and Psychology/Cognitive Sciences. The Sections of the Academy Have Evolved Over the Years and Some Members Elected Long Ago, Such As Robert Yerkes and Karl von Frisch, Are Listed in the Member Directory Without a Section. While the Membership of the United States NAS Is Largely Composed of United States Scientists, Foreign Members Can Be Elected and the List Below Includes Several Non-Americans[a]

Name	General Research Interests	Link to Biographical Information, if Available
Alexander, Richard D	Evolution of social behavior, communication	
Warder Clyde Allee 1885–1955	Animal aggregations, social interactions in flocks	http://www.nasonline.org/publications/biographical-memoirs/memoir-pdfs/allee-warder.pdf
Altmann, Jeanne	Primate behavior	
Beach, Frank 1911–1988	Hormones and behavior	http://www.nasonline.org/publications/biographical-memoirs/memoir-pdfs/beach-frank-a.pdf
Chapman, Frank 1864–1945	Bird behavior	http://www.nasonline.org/publications/biographical-memoirs/memoir-pdfs/chapman-frank-m-1864–1945.pdf
Cheney, Dorothy	Primate behavior	

(*Continued*)

TABLE 5 (Continued)

Name	General Research Interests	Link to Biographical Information, if Available
Cousteau, Jacques-Yves 1910–1997	Underwater exploration, behavior of ocean animals	
de Waal, Frans BM	Primate behavior	
Dethier, Vincent G 1915–1993	Sensory physiology and behavior	http://www.nasonline.org/publications/biographical-memoirs/memoir-pdfs/dethier-vincent.pdf
Dobzhansky, Theodosius 1900–1975	*Drosophila* genetics, including behavioral genetics, evolutionary mechanisms	http://www.nasonline.org/publications/biographical-memoirs/memoir-pdfs/dobzhansky-theodosius.pdf
Eisner, Thomas 1929–2011	Chemoecology, pheromones and chemical defenses	http://www.nasonline.org/publications/biographical-memoirs/memoir-pdfs/eisner-thomas.pdf
Emerson, Alfred E 1896–1976	Termite evolution and behavior, co-author of pioneering ecology textbook	http://www.nasonline.org/publications/biographical-memoirs/memoir-pdfs/emerson-alfred.pdf
Evans, Howard E 1919–2002	Wasp behavior and evolution	http://www.nasonline.org/publications/biographical-memoirs/memoir-pdfs/evans-howard.pdf
Fraenkel, Gottfried S 1901–1984	Sensory physiology	http://www.nasonline.org/publications/biographical-memoirs/memoir-pdfs/fraenkel-gottfried.pdf
Gadagkar, Raghavendra	Social behavior and evolution	
Grant, Peter	Behavior and evolution of Galapagos finches	
Grant, Rosemary	Behavior and evolution of Galapagos finches	
Griffin, Donald 1915–2003	Bat echolocation	http://www.nasonline.org/publications/biographical-memoirs/memoir-pdfs/griffin-donald.pdf
Hall, Jeffrey	*Drosophila* behavioral genetics	
Hinde, Robert B	Ethology, biological bases of human behavior	
Hölldobler, Bert	Ant behavior, communication and evolution	

(*Continued*)

TABLE 5 (Continued)

Name	General Research Interests	Link to Biographical Information, if Available
Hrdy, Sarah B	Primate behavior	
Kandel, Eric R	Learning and memory	http://www.nobelprize.org/nobel_prizes/medicine/laureates/2000/
Krebs, John R	Behavioral ecology	
Kerr, Warwick	Bee behavior	
Lehrman, Daniel S	Comparative psychology, hormones and behavior	Rosenblatt JS. 1995. Daniel Sanford Lehrman June 1, 1919–August 27, 1972. National Academy Biographical Memoirs
Lorenz, Konrad 1903–89	Ethology, imprinting, aggression	http://www.nobelprize.org/nobel_prizes/medicine/laureates/1973/
Macarthur, Robert 1930–72	Population ecology, dispersal behavior, r and K selection	http://www.nasonline.org/publications/biographical-memoirs/memoir-pdfs/mac-arthur-robert-h.pdf
Marder, Eve	Neural circuits and behavior	http://blogs.brandeis.edu/marderlab/research/
Marler, Peter R 1928–2014	Auditory communication	
Maynard Smith, John	Theoretical biology	
Michener, Charles D 1918–2015	Bee evolution and social behavior	http://link.springer.com/article/10.1007%2Fs13592-015-0425-3
Orians, Gordon	Mating systems	
Richter, Curt P 1894–1988	Psychobiology, circadian rhythms	http://www.nasonline.org/publications/biographical-memoirs/memoir-pdfs/richter-curt.pdf
Robinson, Gene E	Neuroscience and molecular genetics of social behavior	
Roeder, Kenneth G 1908–79	Sensory biology	http://www.nasonline.org/publications/biographical-memoirs/memoir-pdfs/roeder-kenneth.pdf

(*Continued*)

TABLE 5 (Continued)

Name	General Research Interests	Link to Biographical Information, if Available
Skinner, BF 1904–90	Comparative psychology, learning	http://www.nasonline.org/publications/biographical-memoirs/memoir-pdfs/skinner-b-f.pdf
Strassmann, Joan	Social evolution, genetics of social systems	
Strier, Karen	Primate behavior	
Tinbergen, Nikolaas 1907–88	Ethology, the four questions	http://www.nobelprize.org/nobel_prizes/medicine/laureates/1973/
		Dewsbury DA. 1990. Nikolaas Tinbergen (1907–88). *American Psychologist* 45:67–68
Tulving, Endel	Experimental psychology, human learning and memory	
Von Frisch, Karl 1886–1982	Honeybee sensory biology, the bee dance language	http://www.nobelprize.org/nobel_prizes/medicine/laureates/1973/
West Eberhard, Mary Jane	Social evolution, mate choice, sexual selection	
Wheeler, William M	Ant evolution and behavior	http://www.nasonline.org/publications/biographical-memoirs/memoir-pdfs/wheeler-william.pdf
Williams, George C 1926–2010	Sexual selection	http://www.nasonline.org/publications/biographical-memoirs/memoir-pdfs/williams-george.pdf
Wilson, Edward O	Ant evolution and behavior, social evolution, communication	
Yerkes, Robert M 1876–1956	Comparative psychobiology, primate behavior	http://www.nasonline.org/publications/biographical-memoirs/memoir-pdfs/yerkes-robert-m.pdf

[a]*It is possible that I have missed relevant Academy members in my search, as section identities and keyword searches in the Academy member directory are not completely effective for the multidisciplinary field of animal behavior.*

TABLE 6 Nobel Prize Winners Whose Work At Least Partly Focused on Animal Behavior

Ivan Pavlov, 1904	https://www.nobelprize.org/nobel_prizes/medicine/laureates/1904/
Thomas H Morgan, 1933	http://www.nobelprize.org/nobel_prizes/medicine/laureates/1933/morgan-bio.html
Konrad Lorenz, 1973	https://www.nobelprize.org/nobel_prizes/medicine/laureates/1973/
Niko Tinbergen, 1973	https://www.nobelprize.org/nobel_prizes/medicine/laureates/1973/
Karl von Frisch, 1973	https://www.nobelprize.org/nobel_prizes/medicine/laureates/1973/
Eric Kandel, 2000	https://www.nobelprize.org/nobel_prizes/medicine/laureates/2000/

In the short essays in this book I strive to capture the history and the future of the study of animal behavior. I start with the distant unrecorded human past, but most of my focus is on the wonderful spirit of discovery that has infused the scientific community of animal behaviorists over the last 200 years. In this small span of time, we have gone from exploring an unknown world to unraveling the mysteries of how genes regulate behavior. Much more remains to be learned about how and why animals behave as they do and this book is more of a snapshot of what we know, rather than a completed portrait of a field of inquiry that is essentially solved. This book represents one person's view on the intellectual history of a very dynamic and complicated field of inquiry and I want to add the caveat that, while I would not expect that all the scientists who work on behavior would embrace my view, I hope that they would at least be intrigued by my decisions.

REFERENCES

Dewsbury, D.A., 1984. Comparative Psychology in the Twentieth Century. Hutchinson Press, New York, 411 pp.

Dewsbury, D.A. (Ed.), 1989. Studying Animal Behavior: Autobiographies of the Founders. University of Chicago Press, Chicago, IL, 512 pp.

Drickamer, L.C., Dewsbury, D.A. (Eds.), 2009. Leaders of Animal Behavior: The Second Generation. Cambridge University Press, Cambridge, UK, 632 pp.

Egerton, F.N., 2016. History of ecological sciences, Part 56: ethology until 1973. Bull Ecol Soc Am 97.

Houck, L.D., Drickamer, L.C. (Eds.), 1996. Foundations of Animal Behavior: Classic Papers With Commentaries. University of Chicago Press, Chicago, IL, 858 pp.

Klopfer, P.H., Hailman, J.P., 1972. Function and Evolution of Behavior. Addison-Wesley, New York, 404 pp.

Wilson, E.O., 1975. Sociobiology: The New Synthesis. Belknap Press of Harvard University Press, Cambridge, MA, 720 pp.

Chapter 1

50,000 Years Before Present: The Dawn of Human Evolution

THE CONCEPT

Human understanding of animal behavior is key to our survival and is representative of the fact that all animals evolve to respond to the behavior of other animals. Human interest in animal behavior extends deep into our evolutionary roots.

THE EXPLANATION

The human study of animal behavior is not new. For our newly minted human ancestors, who lived roughly 50,000 years ago, understanding and reacting to animal behavior was just as important as it is now. In fact, in terms of immediate survival value, knowing about animal behavior was more important for prehistoric humans than it is for present-day humans. Early humans needed to comprehend the behavior of animals that might eat them, as well as of the animals that they wanted to eat. Modern society does much to insulate most humans from the danger of being eaten by an animal, and most contemporary people understand little of the behavior of the fish, fowl, swine, and cattle that form the basis of part of a human omnivore's diet.

Of course the necessity for keen observation of animal behavior predated the evolutionary distinction of humans as a species. You can peel back as far as you like in the family tree of animals and the same necessities will have existed for any species that you might choose. What varies among animal species is how much knowledge of potential predators and food items comes as a result of genetically coded information, versus how much is gained via experience, learning, and calculation. Butterflies and moths inherit their knowledge about what plants to eat, while humans learn what to eat via culture and trial and error.

The best prehistoric agriculturalists learned how to use genetics and evolution to modify the behavior of animal species that humans wanted to exploit. Domesticated versions of horses, sheep, goats, swine, cattle, chickens, ducks, and

Conceptual Breakthroughs in Ethology and Animal Behavior.
DOI: http://dx.doi.org/10.1016/B978-0-12-809265-1.00001-0

rabbits behave very differently than their wild relatives. Using the same principles of selective breeding, humans domesticated cats and dogs as companion animals.

These humans, who did this work before written language existed, learned how to use genetics and evolution by trial and error. Without knowing about chromosomes, genes, or DNA, they engineered major genetic changes in other species. Knowledge of how and why to control breeding of other species came through millennia of doing science without calling the outcomes scientific.

The formal science of animal behavior came into existence much more recently. The exact divide between careful animal husbandry and science is indistinct. The first approaches to animal behavior that seem scientific to us came from two rather distinct streams of human intellectual development. One was the documentation of the natural world that followed the colonization of much of the globe by European nations. In this book, I highlight the work of the naturalists Alexander von Humboldt, John James Audubon, Alfred Russell Wallace, Charles Darwin, Jean-Henri Fabre, and Thomas Belt, all of whom worked in the 1800s, but they are part of the larger fabric of natural exploration that started with the colonization of the Americas in the 1500s (see Chapter 7: 1800s The Great Explorers). Their observations included the behavior of animals in natural settings.

The other avenue was inquiry into the physiological mechanisms that support animal life. William Harvey's discovery of how blood circulated in the early 1600s coincided in time with Charles Butler's (1623; see Chapter 3: 1623 Social Behavior) book on the social behavior of honeybees. Inquiries like Harvey's and Butler's led to deeper experimental explorations of physiology and behavior from the mid-1600s through the 1800s, as well as anatomical studies that supported the understanding of physiology. In addition, Leeuwenhoek developed the microscope as an aid in detailed anatomical studies in the later 1600s, a key advance in biology.

Four elements came together in the 20th century to form the modern biological science of animal behavior:

- Experimental approaches to test for causes of biological phenomena.
- Use of statistics to analyze data from experiments.
- Integration of evolutionary theory with natural history observations.
- An understanding of the physiological underpinnings of behavior.

In the 1970s we entered a period of extensive use of mathematical modeling in behavioral studies. This was coupled with the development of sophisticated evolutionary theory. These changes created a renaissance of thought in ethology and behavioral ecology about how evolution has shaped behavior. In a parallel scientific sphere, comparative psychology integrated with neuroscience, giving new insights into the deeper levels of how the brain regulates behavior. Genomics and its numerous "-omic" offspring, including proteomics, transcriptomics, and metabolomics, link evolution, which drives changes in the genome, with the neuroscience of behavior.

While these fields have not merged into the grand synthesis once envisioned by Wilson (1975) as the outcome of his sociobiological revolution, the current scientific landscape is rich with opportunities for integrating knowledge from these very diverse scientific fields. There is a tension between the need for scientists to be ever-more specialized in order to master increasingly difficult field-specific techniques and the potential benefits that come from seeing the relationships among seemingly isolated scientific findings from very disparate intellectual disciplines.

Behavior lies at the nexus of an animal's interaction with its environment. Humans have changed the global environment, particularly since the beginning of the industrial revolution, in an incomprehensibly rapid and astoundingly thorough manner. We have literally left no stone on the planet unturned. If we are to find ways to preserve the global biota into the next century, knowledge of animal behavior at all levels, ranging from field natural history to the genome, will be needed.

IMPACT: 10

Animal behavior is as important for us as it was for our prehistoric ancestors. The focus of our application of behavioral knowledge has shifted over the millennia from predator avoidance and food discovery, through a phase in which domestication of livestock and companion animals was preeminent, to our current world, in which knowledge of behavior, its evolution, and its regulation, is key to conserving what is left of the natural world.

SEE ALSO

Chapter 2, 12,000 Years Before Present Domestication; Chapter 6, 1800s Birds in Their Natural Setting; Chapter 7, 1800s The Great Explorers.

REFERENCES AND SUGGESTED READING

Most of the chapters in this book have more lists of references and reading suggestions. The possibilities for this chapter are immense, and I can suggest a few books that inspired me:

Belt, T., 1874. The naturalist in Nicaragua. Available in modern reprints, such as the Leopold Classic Library, 344 pp.

Dillard, A., 1974. Pilgrim at Tinker Creek. Harper Collins, p. 304.

Matthiesen, P., 1959. Wildlife in America. Penquin, p. 336.

Schaller, G., 1988. Stones of Silence: Journeys in the Himalayas. Viking, p. 292.

Wiener, J., 1995. The Beak of the Finch. Penguin, p. 332.

Wilson, E.O., 1975. Sociobiology: The New Synthesis. Belknap Press of Harvard University Press, p. 720.

Chapter 2

12,000 Years Before Present: Domestication

THE CONCEPT

Domestication of animals combines applied animal behavior, behavioral genetics, the principles of evolution, and experimentation through trial and error. Humans have applied the principles of these scientific approaches for millennia, only our expression of this as a science is relatively new.

THE EXPLANATION

The history of the dog tells us a lot about human nature, about how we gain scientific knowledge, and about how humans shape and mold nature to fit their needs. The dog is a deeply human construct, and our efforts have given us both a valued companion and deep cultural knowledge of the power artificial selection. Human endeavor took an animal from nature and made it into an extension of human existence. As pointed out in Chapter 1, 50,000 Years Before Present: The Dawn of Human Evolution, humans are intimately tied with their natural surroundings and it is impossible to set a date for the beginning of our awareness of animal behavior, but certainly by the time of the domestication of the dog humans understood much about shaping behavior. The date of the emergence of dogs as a recognizable derivative of wolves is controversial, but the generally accepted timing is 12,000 to 16,000 years ago (Perri, 2016).

The histories of wolves and humans wind around each other, with the two species having a relationship that goes deeper into history than any written record. Approach and avoidance divides wolves and humans, a gap that is bridged by the dog. Dogs evolved from wolves under human stewardship and are not quite identical to their wolf ancestors, as they have a largely co-dependent relationship with their human allies. Dogs were the first evolutionary playground in which humans worked out the strength of genetic family ties in shaping behavior and how human control of the evolutionary process of domestication could yield behaviorally tractable animals. Long before Darwin had pointed out how natural selection

Conceptual Breakthroughs in Ethology and Animal Behavior.
DOI: http://dx.doi.org/10.1016/B978-0-12-809265-1.00002-2

shapes species, humans learned how to use artificial selection to shape plants and animals to meet their needs.

Early humans gained massive amounts of experience with using selective breeding as they shaped the forms and behaviors of dogs. Current evidence shows that dogs, as a recognizable animal form, arose in eastern Asia. During pharaonic times in ancient Egypt, dogs were developed that led to present day breeds. The tesem, a hunting dog, is shown in hieroglyphics, as are other dog breeds. Dogs are referenced in the Bible (*Job* 30:1) and in the Koran (18:18). The domestication of the dog in Asia predates the domestication of sheep in Mesopotamia as well as the domestication of cattle in Africa and the Middle East by at least two millennia (Fig. 2.1).

Dogs spread globally with commerce and human migrations. By the middle ages, dog breeds that are recognizably linked to modern breeds had emerged, as had the general types of breed—working dogs, terriers, retrievers, sight hounds, and so on—that differ so much in their traits. Dogs breed true—matings between spaniels produce spaniels, between dachshunds produce dachshunds—and behavioral traits are as constant within a breed as are physical traits.

The dog's domestication and its transformation into a diverse species which encompasses a huge range of sizes, shapes, and behaviors is the result

FIGURE 2.1 A leashed Babylonian hunting dog, a molosser, probably 6th–8th century BCE. Molossers are the progenitors of the modern day mastiffs. *Maspero, G. History of Egypt, Chaldæa, Syria, Babylonia, and Assyria, vol. 3. London: The Grolier Society, www.gutenberg.net, public domain. Drawn by Faucher-Gudin, from a terra-cotta tablet discovered by Sir H. Rawlinson in the ruins of Babylon, and now in the British Museum.*

of humans learning genetic engineering before scientists developed explanations for genetics, the inheritance of traits, and evolution. At our core, humans are scientists; we instinctively use trial and error to discover the solution to problems and it is no surprise that early humans learned that if they controlled the matings of dogs they could work, over the course of generations, toward such diverse companion and work animals, all molded from a single source.

Grist for the evolutionary mill that helped to shape dogs included the neurochemical framework that reinforces social bonding in both species. Oxytocin and dopamine (see Chapter 27: 1964 Dopamine and Reward Reinforcement) play critical roles in reinforcing social bonds within many mammal species, and in the case of dogs and humans, both species seem particularly tuned to respond to the other with increased levels of oxytocin (e.g., Nagasawa et al., 2009, 2015). Now we know that not only did we shape dogs to fit our needs, but that dogs shaped themselves through evolution to fit into human sociality. This has been a superb outcome for both species.

IMPACT: 8

Modification of behavior through domestication is a key theme in human use of animal behavior in agriculture and in companion animals. This concept, developed far before science existed, informed Darwin's work on natural selection and later studies on behavioral genetics.

SEE ALSO

Chapter 27, 1964 Dopamine and Reward Reinforcement; Chapter 34, 1971 Behavioral Genetics; Chapter 77, 2004 Behavioral Syndromes—Personality in Animals.

REFERENCES AND SUGGESTED READING

Diamond, J., 2002. Evolution, consequences and future of plant and animal domestication. Nature 418, 700–707.

Francis, R.C., 2015. Domesticated: Evolution in a Man-Made World. W. W. Norton & Co, p. 496.

Hare, B., Brown, M., Williamson, C., Tomasello, M., 2002. The domestication of social cognition in dogs. Science 298, 1634–1636.

Nagasawa, M., Kikusui, T., Onaka, T., Ohta, M., 2009. Dog's gaze at its owner increases owner's urinary oxytocin during social interaction. Horm. Behav. 55, 434–441.

Nagasawa, M., Mitsui, S., En, S., Ohtani, N., Ohta, M., Sakuma, Y., et al., 2015. Oxytocin-gaze positive loop and the coevolution of human-dog bonds. Science 348, 333–336.

Perri, A., 2016. A wolf in dog's clothing: initial dog domestication and Pleistocene wolf variation. J. Archaeol. Sci. 68, 1–4.

von Holdt, B.M., Pollinger, J.P., Lohmueller, K.E., et al., 2010. Genome-wide SNP and haplotype analyses reveal a rich history underlying dog domestication. Nature 464, 898–902.

Chapter 3

1623 Social Behavior

THE CONCEPT

The social behavior of another species could be chronicled in a systematic and detailed way, including hypotheses comparing that species' behavior with human behavior.

THE EXPLANATION

Butler's (1623) book on honeybees marked a shift to documentation and analysis of animal behavior that fits within the modern traditions of natural history and science. Much earlier, Aristotle, the 4th century BCE philosopher, had written about natural history. When he was writing, two millennia before Butler's work, Aristotle provided a distinct dividing line in the historical record by providing observation-based accounts of the lives of animals, including honeybees. Prior to that point, natural history was an oral tradition that combined accumulated folk knowledge, theology, and mythology. In the century following Aristotle, and the Roman authors, particularly Cato the Elder, recorded agricultural practices and developed handbooks for Roman agriculture. However, after the work of Aristotle and Cato, there was more reliance on the authority of what had been written than on the development of new knowledge, a pattern that Butler helped to break.

Butler was an accurate and careful observer, and while he did not get everything right about honeybees, he was right often enough to make his book an interesting and valuable source of insight for present day bee researchers. His work can also be seen as a bridge from medieval agrarian traditions to the great naturalists (see Chapter 7: 1800s The Great Explorers) who followed Butler by two centuries.

The scientific interest in honeybees continued in other studies that predated the modern scientific era. Butler's (1623) book was followed by a book by Huber (1806) and then by publications like the *American Bee Journal*, which allowed beekeepers to share information about bees and advice about beekeeping. The Polish beekeeper and scientist, Johan Dzierzon, published work in the mid 1800s that further established the

Conceptual Breakthroughs in Ethology and Animal Behavior.
DOI: http://dx.doi.org/10.1016/B978-0-12-809265-1.00003-4

scientific framework for using honeybees as a model system for studies of social behavior, communication, and sensory physiology. Notably, Dzierzon discovered that honeybee males, or drones, develop from unfertilized eggs.

By the time Karl von Frisch (see Chapter 12: 1914 Sensory Physiology and Behavior) took up the study of bees in the late 1800s Butler, Huber, Dzierzon, the American beekeeper L. L. Langstroth, and others had laid much of the important groundwork for rearing and studying bees. The well known rearing techniques developed by beekeepers and documented by these authors, as well as knowledge about the details of life cycle, diet, and mating made the honeybee an inviting subject for studies of sensory physiology and communication.

IMPACT: 3

Butler was far ahead of his time in his ability to think systematically about what we now would call scientific questions and in his use of an evidence-based approach in finding support for his conclusions. He showed an admirable ability to take what must have been the beekeeping folk wisdom of his time and to condense the facts into this book. His prose is a challenge to modern readers, but his work is so prescient that modern scientists should include Butler on their reading lists.

SEE ALSO

Chapter 9, 1859 Darwin and Social Insects; Chapter 12, 1914 Sensory Physiology and Behavior; Chapter 28, 1964 Inclusive Fitness and the Evolution of Altruism; Chapter 71, 1998 Self-Organization of Social Systems.

REFERENCES AND SUGGESTED READING

Butler, C.G., 1623. The Feminine Monarchie or the Historie of Bees. Full text available online at <http://bees.library.cornell.edu/cgi/t/text/text-idx?c=bees;idno=6371408>.

Dzierzon, J., 1882. Dzierzon's Rational Beekeeping: or the Theory and Practice of Dr. Dzierzon. 2009 reprint, translated into English. Kessinger Publishing, LLC, 366 pp.

Huber, F., 1806. New Observations on the Natural History of Bees. Longman, Hurst, Rees, and Orme, London, p. 300.

Chapter 4

1700s Classifying Life

THE CONCEPT

The lifeforms on earth can be arranged hierarchically, based on degrees of similarity. The highest levels of the hierarchy, microbes, fungi, plants, and animals, represent fundamental roles of life on earth. Species, the lowest level in the scheme, are placed within genera. Species in the same genus typically have structures and behavior that vary only in small details. This concept is particularly valuable in helping us to understand the evolution of behavior.

THE EXPLANATION

Linnaeus taught biologists how to name the living world. By the 1700s, when Linnaeus worked in Sweden, humans already had names for most common plants and animals. Linnaeus created a system that allowed for naming newly discovered plants and animals, as well as for classifying organisms based on their similarities.

The use of Linnaean names allows biologists to collect data on an animal then to be sure that subsequent biologists can find and study the same species. How do names work in biology? A Linnaean name is two words; the first is the genus. A genus includes a set of similar species. Linnaeus had a penchant for using Latin, the intellectual *lingua franca* of his time, so, for example, all of the dozen or so species of honeybees are in the genus *Apis*, the Latin word for bee. The honeybee of the western world, familiar to anyone living in the Americas, Europe, or Africa, is the species *mellifera*, so the full name of our honeybee is *Apis mellifera*. *Mellifera* is the Latin word for sweet, or for honey, so the name makes sense if you know Latin (Fig. 4.1).

Similar genera are grouped into families, similar families into orders, and similar orders into classes. These groupings are hierarchical and are constantly reworked by scientists as we improve our understanding of the living world's evolution. However, it is very important to know that evolution came into the scientific picture after Linneaus' time, and the original intent

Conceptual Breakthroughs in Ethology and Animal Behavior.
DOI: http://dx.doi.org/10.1016/B978-0-12-809265-1.00004-6

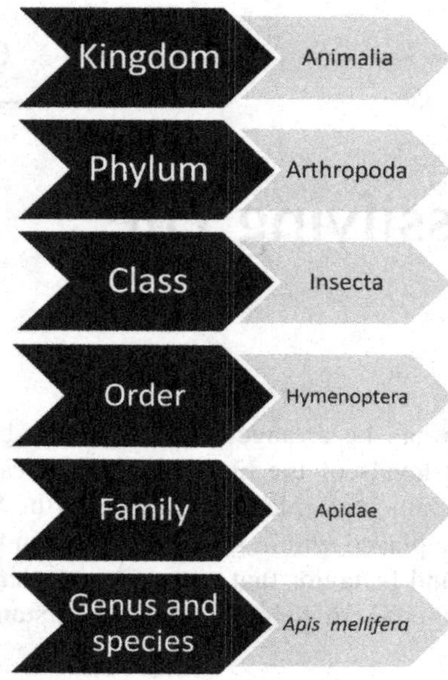

FIGURE 4.1 The hierarchical naming system for animals, with the classification of the honeybee, *Apis mellifera*, as an example.

of the scheme was to classify life, rather than to describe the result of evolution.

Ultimately, the incorporation of evolution into the naming scheme was an even more profound outcome of Linnaeus' naming system than the tools the system gave to scientists for the identification of species. Early scientists believed that a god created an orderly world, in which humans could discover the hierarchy of life. By the 1800s, though scientists had begun to discover that species can change over time, that fossils that exist of extinct plants and animals fit into the hierarchy, and that humans can create species through domestication. The idea that a natural process, evolution, underlies the diversity of the natural world was then incorporated into the classification system.

If species of animals are grouped together in a genus based on their physical characteristics, then why not expect that their behavioral characteristics would be quite similar as well? Behavioral resemblances could also be strong among genera grouped into the same family, and perhaps more weakly so for families placed into the same order. The hierarchy reflects, as you work up the rungs, both decreasing physical similarity and decreasing behavioral similarity.

Physical and behavioral characters equally reverberate through the animal species within a branch of the Linnaean classification. However, dead specimens usually serve as the basis for naming animals. These are often collected by someone other than the taxonomist who makes the classification and also typically are captured hundreds or thousands of kilometers from where the taxonomist works. This creates a fundamental difference between how scientists can discover physical traits versus behavioral traits. Physical traits are present in museum specimens. Behavioral traits are discovered either in notes taken when an animal was collected or by traveling to observe natural populations of the animal.

For physical traits, the classification embodies the differences among animals. In the modern scientific world, knowledge of genetic differences among animals is also brought into the practice of taxonomy. For behavioral traits, the classification is used to establish hypotheses that similarities will be found. This primacy of physical phenotypes in establishing animal classifications means that much behavioral information is lost or ignored in the process. Behavioral dissimilarities among closely related species in the hierarchy suggest interesting evolutionary questions about why those differences appear, including perhaps that the classification is incorrect.

When animal behaviorists of the 20th century began turning their eye to understanding the evolution of behavior, they had a classification based on physical similarities as their starting point. This scientific approach has shifted to the use of tools from molecular genetics to understand how genetic similarity below and above the level of the species correlate with behavior. DNA sequencing has also caused biologists to reconsider important aspects of classifications, which in turn have helped animal behaviorists to understand the intricacies of behavioral evolution.

IMPACT: 7

The use of the hierarchical classification to describe biological diversity and to visualize the results of the evolutionary process is a keystone of modern biology. This information is used in behavioral studies. In addition to testing evolutionary hypotheses, Linnaean classification gives biologists a naming system that helps to prevent confusion about the identity of animals under study.

REFERENCES AND SUGGESTED READING

Blunt, W., 2002. The Compleat Naturalist: A Life of Linnaeus. Frances Lincoln Ltd. (originally published in 1971 by Harper Collins), London, p. 288.

Kitching, I.J., Forey, P.L., Humphries, C.J., Williams, D.M., 1998. Cladistics: The Theory and Practice of Parsimony Analysis, second ed. Oxford University Press, p. 228.

Koerner, L., 2001. Linnaeus: Nature and Nation. Harvard University Press, p. 298.

Williams, D., Schmitt, M., 2016. The Future of Phylogenetic Systematics: The Legacy of Willi Hennig. Cambridge University Press, p. 508.

Chapter 5

1729 Biological Clocks

THE CONCEPT

Internal timekeepers in organisms synchronize internal functions and keep the timing of those functions correct with respect to the daily cycle of light and dark.

THE EXPLANATION

This is the only chapter in this animal behavior book in which plants play a leading role. In 1729 Jean-Jacques d'Ortous de Mairan truly broke ground with his simple experiment with a plant that opens and closes its leaves on a daily cycle. This is common in plants in the mimosa family—the leaves spread and appear to expand as the sun rises and then fold as the sun sets.

If the plant could sense the sun's presence, this would be the obvious mechanistic explanation for the triggering of the leaf movements. De Marain, wondered, though, what would happen if the plant were kept in the dark, so that the influence of the sun was absent? The result of this experiment was that the plant maintained its rhythm of leaf movements even when the stimulus of the sun was removed.

This suggested that the plant had another way of synchronizing its movements with the sun. De Marain was an astronomer and never personally followed up on this experiment but his short publication served as a stimulus for other scientists to pick up this thread of investigation. Ultimately this work led to the discovery of biological clocks, the timepieces that keep biological systems in tune with daily and seasonal fluctuations of the environment. Biological clocks exist in nearly all organisms, including microbes, fungi, plants, and animals.

Obviously, we now know quite a lot more about biological clocks, starting with Aschoff's (1960) explorations of circadian rhythms. Biological clocks regulate everything from digestion to awareness. Animals have multiple biological clocks, and how they keep those clocks in time with the surrounding

Conceptual Breakthroughs in Ethology and Animal Behavior.
DOI: http://dx.doi.org/10.1016/B978-0-12-809265-1.00005-8

world and with each other is a marvel of physiology. Chapter 22, 1954 Zeitgebers (Time-Givers) for Biological Clocks, deals with how animals set their clocks.

IMPACT: 3

Biological clocks are fundamental to the functioning of life and to the organization and coordination of behavior. Simple behavioral functions, such as timing active and inactive periods during the day/night cycle to maximize productivity and minimize risk rely on internal clock functions. The early recognition by de Marain of the existence and functioning of biological clocks, followed by Aschoff's key investigations in the mid-1900s, has led to knowledge of the genetics underlying biological clocks, into how they are set, and into how clocks regulate physiological and behavioral processes.

SEE ALSO

Chapter 22, 1954 Zeitgebers (Time-Givers) for Biological Clocks; Chapter 25, 1960 Motivation and Drive.

REFERENCES AND SUGGESTED READING

Anon, 2008. Clock classics: It all started with the plants. <http://scienceblogs.com/clock/2008/05/29/clock-classics-it-all-started/>.

Aschoff, J., 1960. Exogenous and endogenous components in circadian rhythms. Cold Spring Harb. Symp. Quant. Biol. 25, 11–28.

De Mairan, J.J.O., 1729. Observation Botanique, Histoire de l'Academie Royale des Sciences, Paris, p. 35. <http://galileo.rice.edu/Catalog/NewFiles/mairan.html>.

King, D.P., Takahashi, J.S., 2000. Molecular genetics of circadian rhythms in mammals. Annu. Rev. Neurosci. 23, 713–742.

Lowrey, P.L., Takahashi, J.S., 2000. Genetics of the mammalian circadian system: photic entrainment, circadian pacemaker mechanisms, and post-translational control. Annu. Rev. Genet. 34, 533–562.

Pittendrigh, C.S., 1960. Circadian rhythms and the circadian organization of living systems. Cold Spring Harb. Symp. Quant. Biol. 25, 159–184.

Wager-Smith, K., Kay, S.A., 2000. Circadian rhythm genetics: from flies to mice to humans. Nat. Genet. 26, 23–27.

Chapter 6

1800s Birds in Their Natural Setting

THE CONCEPT

Art and biology can come together to increase our understanding of animal behavior.

THE EXPLANATION

Art, science, and animal behavior have interacted strongly since humans first recorded scenes on cave walls and cliff faces. In art, animals appear as companions to Greek gods in ancient art, in gold pieces produced in Central and South America before 1500, and as props and scenic devices in landscape and still-life paintings from many eras.

John James Audubon represents a transition in bringing precision of representation of animal behavior to his paintings. Born in 1785 in Haiti, he lived first in France and then, once he reached the age of 18, in the United States. He was obsessed early in life with painting but only turned to painting as a profession after a short career as a businessman. His *Birds of America* contains classic representations of nearly all of the birds found in America north of Mexico. He died in 1851.

Key to Audubon's influence was his determination to represent birds as they appeared in life, poised, rather than posed, for their next action. Audubon inspired generations of ornithologists starting with George B. Grinnell, founder of the Audubon Society and a force in the establishment of America's national parks. He was a contemporary of Alexander Wilson (1766–1813), who established the science of ornithology in the Americas. Audubon's life-like portraits of birds opened the window for responses about behavior, in addition to the esthetics of color and form.

Conceptual Breakthroughs in Ethology and Animal Behavior.
DOI: http://dx.doi.org/10.1016/B978-0-12-809265-1.00006-X

FIGURE 6.1 John James Aubudon as a young man, from a portrait by Henry Inman. *From Audubon, M.R., 1897. Aubudon and His Journals. Charles Scribner's Sons, New York.*

Audubon wrote exquisitely about the development of his artistic skills in an essay "My style of drawing birds," which is reprinted in Audubon and Audubon (1986). He says that from early in life:

nothing could ever answer my enthusiastic desires to represent nature, except to copy her in her own way, alive and moving!

p. 523

Audubon became skilled in mounting dead birds (he was apparently quite a good shot) using wires to suspend the corpses in life-like poses (Fig. 6.1).

As I wandered, mostly bent on the study of birds, and a wish to represent all those found in our woods, to the best of my powers, I gradually became acquainted with their forms and habits, and the use of my wires was improved by constant practice...after a time I laid down what I was pleased to call a constitution of my manner of drawing birds, formed upon natural principles...

pp. 525–526

Audubon's impact extended through the work of 20th century artists' paintings of birds and also probably informs the desires of photographers and videographers to obtain lifelike images. Thoughtful viewing of his work is still inspirational and each new generation of ornithologists and animal behaviorists finds at least part of their inspiration in Audubon's revolutionary paintings (Fig. 6.2).

FIGURE 6.2 Audubon's illustration of the American Cormorant. *Source: Plate 266 of Birds of America by John James Audubon depicting Common Cormorant, 1827–1838, John James Audubon (1785–1851), University of Pittsburgh.*

IMPACT: 6

Audubon inspired generations of interest in birds and in nature, including of course the society of birders that carries his name. He led the way in representing animal behavior naturalistically, giving life to forms in ways that allow the viewer to visualize behavior as well as ornamentation. Like the great explorers (see chapter: 1800s The Great Explorers) he helped build the foundations of natural history that fuel modern biology.

REFERENCES AND SUGGESTED READING

Audubon, J.J., Audubon, M.R., 1986. Audubon and His Journals, vol. 2. Dover Publications, Inc, 554 pp. Also available at: <https://books.google.com/books?id=zgWDBj7qnz0C>.

Chapter 7

1800s The Great Explorers

THE CONCEPT

In the 18th and 19th centuries, the earth beyond England, Europe, and the Orient contained unimaginable biological diversity that merited exploration and documentation. Much that attracted the explorers are related to animal behavior.

THE EXPLANATION

The 1700s and 1800s brought an intense interest in the natural. Global conquest was nearly complete and the interest of both governments and the public turned to learning about the contents of their realms. Many explorers risked life and limb in quests to create maps and to document the geography and natural history of continents and archipelagos that were new in the eyes of Europeans. Explorers wrote accounts of their journeys that were popular and literary successes. They also brought back geological and biological collections that formed the basis for generations of scientific study in the great museums of Europe.

The explorers showed us that the wider world was not just a setting for commerce and conquest, but also a world of natural wonder with beauty, diversity, and meaning far beyond their imagination. Among the most influential were:

Alexander von Humboldt (1769–1859), who explored Mexico, Central and South America from 1799 to 1804 and recorded his observations and interpretations of the natural world in a book called the *Kosmos*.

Meriwether Lewis (1774–1809) and *William Clark* (1770–1838) were tasked by President Thomas Jefferson with mapping and documenting the American West. Their epic journey, from 1804 to 1806, into the upper reaches of the Mississippi river drainage and then across ultimately to the Columbia River and the Pacific Ocean, captured the attention of the young nation.

Charles Darwin (1809–82) served as a naturalist aboard the British naval ship the *Beagle* from 1831 to 1836 and published an account of his circumnavigation of the globe in *The Voyage of the Beagle*. Darwin's observations while

Conceptual Breakthroughs in Ethology and Animal Behavior.
DOI: http://dx.doi.org/10.1016/B978-0-12-809265-1.00007-1

FIGURE 7.1 Henry Bates illustrated his *The Naturalist on the River Amazons* with perceptive drawings of plants, animals, and people. Remembering that the books of the great naturalists were their first window into the largely unknown world outside of Europe and England, these drawings gave visual support to the written descriptions. Here is Bates' representation of the fabulous beak of a toucan.

on the Beagle formed the basis for his numerous scientific contributions, including his development of evolutionary theory (see Chapter 8: 1859 Darwin and Behavior). Darwin's natural history accounts helped to focus public attention on plant and animal diversity beyond the confines of England and Europe.

Alfred Russel Wallace explored the Amazon region of Brazil from 1848 to 1852 and the Malay Archipelago from 1854 to 1862. He suffered one of the most famous mishaps of the history of exploration when his collections and notes from Brazil were lost, in their entirety, when the ship carrying them burned on the voyage home. His book, *The Malay Archipelago*, opened the eyes of the public to a previously nearly completely unknown region of the world. Many credit Wallace, concurrently with Charles Darwin, with realizing the importance of natural selection and evolution.

Henry Walter Bates (1825–92) accompanied Wallace on part of the Amazon expedition and wrote the classic *The Naturalist on the River Amazons* (Figs. 7.1 and 7.2).

Thomas Belt (1832–78) trained as a mining engineer, pursued natural history as a pastime while working on mining projects. His *The Naturalist in Nicaragua* is a passionate and consuming account of Central American natural history, with fascinating accounts of crocodiles, and the relationship between bullhorn acacias and the ants that occupy their thorns.

IMPACT: 8

These explorers opened the natural world to scientists' eyes, documented landscapes that humans have now changed beyond recognition, and set the

Scarlet-faced and Parauacú Monkeys

FIGURE 7.2 In this drawing, Bates shows two species of monkey. The detailed and expressive drawing style helps transport the reader, in their imagination, to the Amazon.

table for modern evolutionary theory. Science owes them a large debt of gratitude for their dangerous and challenging work.

REFERENCES AND SUGGESTED READING

Ambrose, S.E., 1997. Undaunted Courage: Meriwether Lewis, Thomas Jefferson, and the Opening of the American West. Simon and Schuster, 521 pp.

Bates, H.W., 1863. The Naturalist on the River Amazons. Available in reprint versions from various publishers.

Belt, T., 1874. The Naturalist in Nicaragua. Available in modern reprints, such as the Leopold Classic Library, 344 pp.

Darwin, C.R., 1839. The Voyage of the Beagle. John Murray, London, Many modern editions available. Ebook: <http://www.gutenberg.org/ebooks/944 http://literature.org/authors/darwin-charles/the-voyage-of-the-beagle/>.

Peattie, D.C., 1936. Green Laurels: The Lives and Achievements of the Great Naturalists. Simon and Schuster, New York.

von Humboldt, A., 1850. Views of Nature, or Contemplations on the Sublime Phenomena of Creation. Ebook: <https://archive.org/details/viewsnatureorco00bohngoog>.

Wallace A.R. 1869. The Malay Archipelago: The Land of the Orangutan, and the Bird of Paradise. A Narrative of Travel, With Sketches of Man and Nature. Ebook: <https://ebooks.adelaide.edu.au/w/wallace/alfred_russel/malay/>.

Chapter 8

1859 Darwin and Behavior

THE CONCEPT

Natural selection shapes behavior, and sexual selection has a special role in the evolution of signals used in mating and courtship. Domestication is the result of selection that is guided by humans. Animals have emotions, which are expressed in their social interactions.

THE EXPLANATION

Modern biology began with Charles Darwin, and it's not surprising that his work stands as the foundation for contemporary thinking about animal behavior. Darwin's work on *natural selection, sexual selection, domestication*, and *animal emotions* flavors all current research on animal behavior. Darwin speaks best in his own voice, so this chapter largely relies on four quotations from Darwin, with some comments from me (Fig. 8.1).

Natural selection: Essentially, this is the survival of the fittest, but its important to remember that to an evolutionary biologist, fitness is measured in the number of offspring an animal has. In the evolutionary world surviving means nothing without offspring; the winnowing of characteristics done by evolution favors features that improve survival *and* reproduction.

One general law, leading to the advancement of all organic beings, namely, multiply, vary, let the strongest live and the weakest die.

The Origin of Species

Sexual selection: Traits that help animals to find mates are particularly favored by selection, and anything that makes an animal more attractive to potential mates confers an advantage. Selection of secondary sexual characteristics gained Darwin's attention and led to his discussion of sexually selected characteristics. We now know, of course, that sexual selection works as strongly on females as on males.

Sexual selection is, therefore, less rigorous than natural selection. Generally, the most vigorous males, those which are best fitted for their places in nature, will leave

Conceptual Breakthroughs in Ethology and Animal Behavior.
DOI: http://dx.doi.org/10.1016/B978-0-12-809265-1.00008-3

FIGURE 8.1 Charles Darwin, painted in 1875 by Walter William Ouless. Public domain.

most progeny. But in many cases victory depends not so much on general vigour, but on having special weapons, confined to the male sex. A hornless stag or spurless cock would have a poor chance of leaving numerous offspring. Sexual selection, by always allowing the victor to breed, might surely give indomitable courage, length of spur, and strength to the wing to strike in the spurred leg, in nearly the same manner as does the brutal cockfighter by the careful selection of his best cocks.

The Origin of Species

Domestication: Humans intervene in the evolutionary process and arrange matings for domesticated species, thus shaping the form and behavior of species like dogs, cats, horses, cattle, and sheep. Domesticated animals tend to be more docile and more easily socialized than their wild counterparts. Ironically, domestication typically favors animals that are unable to foil their human handlers; they are perhaps less well equipped than their wild counterparts to solve problems like gate latches.

How strongly these domestic instincts, habits, and dispositions are inherited, and how curiously they become mingled, is well shown when different breeds of dogs are crossed. Thus it is known that a cross with a bull-dog has affected for many generations the courage and obstinacy of greyhounds; and a cross with a greyhound has given to a whole family of shepherd-dogs a tendency to hunt hares. These domestic instincts, when thus tested by crossing, resemble natural instincts, which in a like manner become curiously blended together, and for a long period exhibit traces of the instincts of either parent...

The Origin of Species

Emotion in animals: The strong, and appropriate, move in science to avoid anthropomorphism caused animal behaviorists to shy away from Darwin's writings on animal emotions for over a century. In the last two decades, scientific discussions of animal emotions such as joy, pleasure, anger, fear, grief, obsession, and depression have become more acceptable. All of these seem to be within the reach of many of our mammalian kindred. It is interesting that spite, which is a more complex mix of emotion and calculation, is perhaps restricted to humans and just a few of our most closely related primate species.

In these cases of the dog and cat, there is every reason to believe that the gestures both of hostility and affection are innate or inherited; for they are almost identically the same in the different races of the species, and in all the individuals of the same race, both young and old.

Darwin (1872)

IMPACT: 10

Google Scholar credits Charles Darwin with over 100,000 citations and he could easily be cited by every scholarly study in evolution or ecology. With more than 150 years having passed since publication of the *Origin of Species*, Darwin's impact is massive.

SEE ALSO

Chapter 2, 12,000 Years Before Present Domestication; Chapter 34, 1971 Behavioral Genetics; Chapter 38, 1973 Game Theory; Chapter 40, The Red Queen.

REFERENCES AND SUGGESTED READING

Darwin, C.R., 1839. The Voyage of the Beagle. John Murray, London. Many modern editions available. Ebook: <http://www.gutenberg.org/ebooks/944>, <http://literature.org/authors/darwin-charles/the-voyage-of-the-beagle/>.

Darwin, C.R., 1859. On the Origin of Species. John Murray, London. Many modern editions are available. Ebook: <http://www.gutenberg.org/ebooks/2009>.

Darwin, C.R., 1872. The Expression of the Emotions in Man and Animals. John Murray, London. Many modern editions are available. Ebook: <http://www.gutenberg.org/ebooks/1227>.

Chapter 9

1859 Darwin and Social Insects

THE CONCEPT

Darwin recognized that the sterile castes of social insects presented a small problem for his theory of natural selection, but that the evolution of different forms within the sterile castes was more difficult to explain.

THE EXPLANATION

Darwin had some difficulty with using natural selection in explaining social insect castes. This extensive quote from *The Origin of Species* explains his concern:

The subject well deserves to be discussed at great length, but I will here take only a single case, that of working or sterile ants. How the workers have been rendered sterile is a difficulty; but not much greater than that of any other striking modification of structure; for it can be shown that some insects and other articulate animals in a state of nature occasionally become sterile; and if such insects had been social, and it had been profitable to the community that a number should have been annually born capable of work, but incapable of procreation, I can see no especial difficulty in this having been effected through natural selection. But I must pass over this preliminary difficulty. The great difficulty lies in the working ants differing widely from both the males and the fertile females in structure, as in the shape of the thorax, and in being destitute of wings and sometimes of eyes, and in instinct. ... If a working ant or other neuter insect had been an ordinary animal, I should have unhesitatingly assumed that all its characters had been slowly acquired through natural selection; namely, by individuals having been born with slight profitable modifications, which were inherited by the offspring, and that these again varied and again were selected, and so onwards. But with the working ant we have an insect differing greatly from its parents, yet absolutely sterile; so that it could never have transmitted successively acquired modifications of structure or instinct to its progeny. It may well be asked how it is possible to reconcile this case with the theory of natural selection?

The Origin of Species

Conceptual Breakthroughs in Ethology and Animal Behavior.
DOI: http://dx.doi.org/10.1016/B978-0-12-809265-1.00009-5

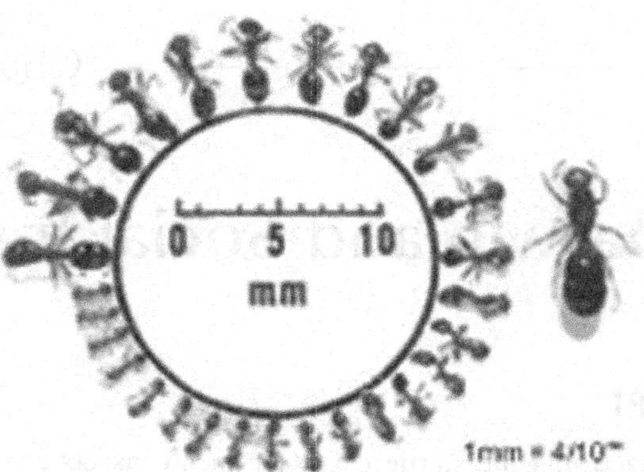

FIGURE 9.1 Caste in the fire ant. The animals arranged in the circle are all sterile workers from the same colony. Workers of different size engage in different tasks. Because workers are sterile, Darwin saw difficulties in explaining such variation using evolutionary principles. The larger ant to the right of the circle is a queen. Photo: Sanford Porter, USDA-ARS, public domain.

In reading the quote it is most important to notice what Darwin was not concerned about: The evolution of sterile castes ("I can see no especial difficulty in this having been effected through natural selection"). Darwin could easily see that sterile workers whose labor benefitted their mother could be explained by the passage of the mother's genes into the next reproductive generation. As Ratnieks et al. (2011) point out in a very insightful analysis, genetics as a field was not born when Darwin was developing his ideas, so he could not have worked out the details of how this came about, but he had a clear view of how sterile castes might have evolved.

In fact Darwin's great difficulty was in explaining how "working ants differing widely from both the males and the fertile females in structure" come to exist. A modern understanding of genetics and development helps us to see that very different external structures can be built in animals that have the genes. An equally modern understanding of evolution sees the workers, in all their forms and variants, as phenotypic extensions of the queen. If queens produce societies that are more successful, then the genes that result in this success are favored by selection.[1]

Unfortunately, Darwin is often misquoted to the effect that his "great difficulty" in explaining social insects relates to the evolution of worker sterility or to the general phenomenon of self-sacrifice (altruism), a misimpression that has been difficult to dispel (Fig. 9.1).

1. This incorrect representation of Darwin's view is a point that creationists have tended to latch onto in attempts to build an argument that even Darwin thought that evolution doesn't work.

IMPACT: 4

Darwin's thoughts about the evolution of the worker castes in social insects informed and stimulated subsequent work by major figures in social insect research, such as William Morton Wheeler, Alfred E. Emerson, Charles D. Michener, and Edward O. Wilson.

SEE ALSO

Chapter 28, 1964 Inclusive Fitness and the Evolution of Altruism; Chapter 35, 1971 Reciprocal Altruism; Chapter 45, 1975 Group Selection; Chapter 46, 1975 Sociobiology.

REFERENCES AND SUGGESTED READING

Breed, M.D., 2016. Charles Michener, 1918–2015. Apidologie. Available from: http://dx.doi.org/10.1007/s13592-015-0425-3.

Darwin, C.R., 1859. On the Origin of Species. John Murray, London. Many modern editions are available. Ebook: <http://www.gutenberg.org/ebooks/2009>.

Emerson, A.E., 1937. Termite nests—a study of the phylogeny of behavior. Science 85, 56.

Michener, C.D., 1974. The Social Behavior of the Bees. Belknap Press of Harvard University Press, 418 pp.

Parker, G.H., 1938. William Morton Wheeler. National Academy of Sciences Biographical Memoirs, <http://www.nasonline.org/publications/biographical-memoirs/memoir-pdfs/wheeler-william.pdf>.

Ratnieks, F.L.W., Foster, K.R., Wenseleers, T., 2011. Darwin's special difficulty: the evolution of "neuter insects" and current theory. Behav. Ecol. Sociobiol. 65, 481–492. Available from: http://dx.doi.org/10.1007/s00265-010-1124-8.

Wheeler, W.M., 1910. Ants; Their Structure, Development and Behavior. Columbia University Press, New York. Ebook: <https://archive.org/details/antstheirstruct00wheegoog>.

Wilson, E.O., 1971. The Insect Societies. Belknap Press of Harvard University Press, Cambridge, Massachusetts, 562 pp.

Wilson, E.O., Michener, C.D., 1982. Alfred Edwards Emerson. National Academy of Sciences Biographical Memoirs, <http://www.nasonline.org/publications/biographical-memoirs/memoir-pdfs/emerson-alfred.pdf>.

Chapter 10

1882 George Romanes and the Birth of Comparative Psychology

THE CONCEPT

Comparing animal and human behavior can yield productive insights into the lives of all animals.

THE EXPLANATION

Comparative psychology is the exploration of animal behavior using the techniques and tools of psychology, with a focus on human–animal comparisons. George Romanes was the leading intellectual force behind the establishment of comparative psychology as an academic discipline in the late 19th century.

This field had a parallel course of development with naturalistic studies of animal behavior that became the focus of ethology. The differences between comparative psychology and ethology were not as stark as they became later, in the 1940s and 1950s, and Romanes had a keen interest in the behavior of animals in natural settings and in evolution.

Romanes made a detailed study of animal intelligence (Romanes, 1882), noting in the preface that:

My second, and much more important object, is that considering the facts of animal intelligence in their relation to the theory of Descent. With the exception of Mr. Darwin's admirable chapters on the mental powers and moral sense, and Mr. Spencer's great work on the Principles of Psychology, there has hitherto been no earnest attempt at tracing the principles which have been probably concerned in the genesis of the Mind.

Animal Intelligence, preface, p. vi

This nod to Darwin had huge importance in bringing evolutionary thinking to the study of animal behavior at an early date. Romanes' focus on evolution was key to the development of comparative psychology as a field.

Conceptual Breakthroughs in Ethology and Animal Behavior.
DOI: http://dx.doi.org/10.1016/B978-0-12-809265-1.00010-1

Romanes spent his academic career in England, but comparative psychology took root in the United States, with leading lights like J. B. Watson, E. L. Thorndike, B. F. Skinner (see Chapter 13: 1938 Skinner and Learning), T. C. Schneirla, and Daniel Lehrman (see Chapter 20: 1953 The Chasm Between Ethology and Comparative Psychology) continuing Romanes' legacy.

Comparative psychology became a subject of some scorn among ethologists and behavioral ecologists in the 1970s. Dewsbury (1984) captures the view from that era "The image of comparative psychology is that of a human-oriented, laboratory-based, nonevolutionary study of trivial behavioral patterns in a few domesticated species" (p. 1). While Dewsbury (1984) vehemently argues against this view in his book, the stain of this image persisted until the 2000s, by which time neuroscience had displaced comparative psychology in most academic psychology departments. Ironically, neuroscientists tend to call on an even more limited range of study animals than did comparative psychologists.

IMPACT: 8

Romanes' impact is felt through the intellectual history of comparative psychology and the integration that came later of ethology and comparative psychology. Contemporary students in comparative psychology and neuroscience are not often asked to read his work or consider his ideas, as he is relegated to a brief mention in an introductory chapter of introductory psychology textbooks. But it is worthwhile to pause a moment and recognize that he propelled an infant field at the critical moment in its development. This is also a good point to consider that scientists who are giants in their field often recede in recognition over time, as the edifice of science is built on their foundation, but the foundation is shielded from view by all that comes later. In this respect Darwin is quite exceptional, as he has remained in the view of both scholars and the general public.

SEE ALSO

Chapter 11, 1894 Morgan's Canon; Chapter 20, 1953 The Chasm Between Ethology and Comparative Psychology; Chapter 26, The Four Questions.

REFERENCES AND SUGGESTED READING

Dewsbury, D.A., 1984. Comparative Psychology in the Twentieth Century. Hutchinson Press, New York, 411 pp.

Romanes, G., 1882. Animal Intelligence. Kegan Paul, Trench, London, Ebook: <https://archive.org/details/animalintelligen00romauoft>.

Thorndike, E.L., 1907. The Elements of Psychology. A.G. Seiler, New York, 351. pp. Ebook: <https://babel.hathitrust.org/cgi/pt?id=loc.ark:/13960/t3zs3kj12;view=1up;seq=11>.

Watson, J.B, 1913. Psychology as the behaviorist views it.. Psychological Review 20, 158–177.

Chapter 11

1894 Morgan's Canon

THE CONCEPT

Do not over-credit animals with human-like capacities, look for the simplest possible explanations for animal behavior.

THE EXPLANATION

Morgan's canon tells us to always go for the simplest possible explanation for animal behavior: It is the Occam's razor for behavioral studies. Morgan's canon has been at once a blessing and a curse for the science of animal behavior. It is a blessing because it stands guard against simply making up stuff about feelings and motivations that animals would have if their brains were perfect mirrors of the human brain. It is a curse because Morgan's canon has stood as a wall against appropriately crediting animals for cognitive abilities, pain perception, complex motivations, and specific strategies such as deceit and spite.

C. L. Morgan (1894) stated:

in no case may we interpret an action as the outcome of a higher psychical faculty, if it can be interpreted as the outcome of the exercise of one that stands lower in the psychological scale (Morgan, 1894, p. 53). Quoted by Elwood R. W. and Arnott, G. 2012. Understanding how animals fight with Lloyd Morgan's canon. Animal Behaviour *84: 1095–1102.*

At heart, Morgan's canon is a strong statement against anthropomorphism. Most contemporary biology students are schooled to expunge anthropomorphic thinking from their scientific toolkits. There is considerable value in this when teaching biology, particularly animal behavior. Morgan's canon helps to overcome the fact that we grow up watching animated feature films and cartoons in which all sorts of animals speak and think like humans. It is fully understandable that many of us arrive at our interest in animal behavior because we have been attracted to animated characters which display empathy, compassion, cunning, perhaps evil, and very deep abilities to plan ahead.

Conceptual Breakthroughs in Ethology and Animal Behavior.
DOI: http://dx.doi.org/10.1016/B978-0-12-809265-1.00011-3

Morgan's canon protects against believing, without evidence, that animals have these behavioral and cognitive properties. Even if we wonder if at least some animals should be credited with these characteristics, we can treat these thoughts as hypotheses to be tested, not abilities to be assumed or imagined.

Yet, we can go quite wrong if we enter the study of animal behavior with the equally uncritical view that animals are unthinking and unfeeling automatons. Portrayals of animals in this way can lead to unthinkable cruelties and unpardonable callousness. It has taken scientists who study animal behavior much too long to recognize that cognition, emotion, and pain represent testable hypotheses, and that the animals under study have been underserved if we dogmatically believe that all animals lack these capabilities.

Donald Griffin, whose discoveries about bat echolocation are highlighted in Chapter 15, 1941 Bat Echolocation, was an early advocate (in the 1990s) of giving more credit to animals. One of his core arguments was that anecdote actually has scientific power; when an animal is observed doing something remarkable, the fact shouldn't be discredited because the behavior was seen only once. Griffin, despite the fact that he was a leading scientist in the mid-20th century, got little traction with mainstream thought when he made this argument:

Ethologists and comparative psychologists have found that the social behavior, discrimination learning, and especially the communicative behavior of many animals are sufficiently versatile to call into question the customary denial that animals have mental experiences comparable to our own. Many of their behavior patterns suggest that animals have mental images of objects, events, or relationships remote from the immediate stimulus situation, as well as self-awareness and intentions concerning future actions. Our behavioristic Zeitgeist has inhibited investigation of such possibilities, but reopening these long-neglected questions requires no departure from the materialistic approach to biological and behavioral science.

Griffin (1978)

It really has been only in the last 20 years that consideration of animal cognition, thoughts and feelings has gained substantial scientific credibility.

There is a thin line to walk between under- and overcrediting the capacities of animals. Sometimes assertions made by people who view themselves as animal advocates are truly cringeworthy. Scientists, or really anyone interested in discovery, rational discourse, and human welfare, should be aware of the types of arguments made by zealots and have counterarguments at hand.

We can acknowledge that at times scientists seem to be blinded by their training to the consideration of reasonable and worthwhile hypotheses about animal cognition and emotions. As in many areas of life, an open, inquiring mind probably leads science to a closer approximation of the truth. The present-day viability of Morgan's canon lies not in denying what animals might do, but rather in encouraging healthy scientific skepticism that leads to well-designed tests of what animals actually do.

IMPACT: 8

Morgan's canon fueled a century of opposition to anthropomorphic thinking in science, which, in large part, was more of a help than a hindrance in the development of the science of animal behavior. It is now time to think critically about when Morgan's canon applies and when it limits thinking in unproductive ways.

SEE ALSO

Chapter 10, 1882 George Romanes and the Birth of Comparative Psychology; Chapter 20, 1953 The Chasm Between Ethology and Comparative Psychology; Chapter 26, The Four Questions.

REFERENCES AND SUGGESTED READING

Bekoff, M., 2008. The Emotional Lives of Animals: A Leading Scientist Explores Animal Joy, Sorrow, and Empathy—and Why They Matter. New World Library, 240 pp.

Griffin, D.R., 1978. Prospects for a cognitive ethology. Behav. Brain Sci. 1, 527–538.

Griffin, D.R., 2001. Animal Minds: Beyond Cognition to Consciousness. University of Chicago Press, Chicago, 376 pp.

Morgan, C.L., 1894. An introduction to Comparative Psychology. W. Scott, London, 438 pp.

Singer, P., 2005. In Defense of Animals: The Second Wave. Wiley, 264 pp.

Waters, R.H., 1939. Morgan's canon and anthropomorphism. Psychol. Rev. 46, 534–540, <http://0-dx.doi.org.libraries.colorado.edu/10.1037/h0055191>.

Chapter 12

1914 Sensory Physiology and Behavior

THE CONCEPT

Contrary to a common belief at the time of this study, many animals, including honeybees, can discriminate between colors.

THE EXPLANATION

One of the persistent mythologies about the divide between humans and other animals was the belief that only humans held the capability for seeing colors. Karl von Frisch, an Austrian bee scientist, used a simple and elegant experiment to test bees for color vision (von Frisch 1914). He conclusively determined that honeybees do, indeed, have color vision.

This finding led to tests in many different animals for abilities to discriminate among colors. Among the many fascinating findings are that some animals can see colors that are not part of the human color range, such as ultraviolet and far red. Humans have trichromatic vision, which involves three primary colors. Some animals are dichromatic, seeing two primary colors, while others see four or more primary colors. Humans blend perception of colors that lie between our primary colors, while some animals handle intermediate colors much differently. A collection of papers honoring the hundredth anniversary of von Frisch's discovery (Dyer and Arikawa, 2014) highlights the diversity of animal visual perception.

Generally color vision has evolved in animals whose niche makes color discriminations appropriate. Activity in daylight is key, of course, and color vision seems to have evolved for within-species and between-species communication systems involving visual signals. Color signals can easily be affected by sexual selection that drives the evolution of much of the signaling seen with species (see Chapter 8: 1859 Darwin and Behavior). While we cannot know conclusively that flower colors evolved as attractants for pollinators like bees, and that color discrimination evolved to aid pollinators

Conceptual Breakthroughs in Ethology and Animal Behavior.
DOI: http://dx.doi.org/10.1016/B978-0-12-809265-1.00012-5

FIGURE 12.1 A honeybee foraging. Bees rely on color vision to find flowers and for navigation. *Courtesy Photo: Michael Breed.*

in finding flowers, these hypotheses are good guesses at how color vision developed in interspecific communication (Fig. 12.1).

von Frisch was ultimately awarded the 1973 Nobel Prize for Physiology or Medicine for his discoveries about bee behavior and for his role in establishing the field of ethology. This prize was shared with Nikolaas Tinbergen (see Chapter 26: The Four Questions) and Konrad Lorenz. Each of these scientists made unique and important discoveries; von Frisch, as the older member of the group, set a tone of meticulous scientific investigation featuring rigorously controlled experiments. In addition to the discovery of color vision by bees, highlighted in this chapter, von Frisch is celebrated for his discovery of the dance language of the bees.

IMPACT: 6

Assuming that animals have limited sensory abilities that are inferior to those of humans was one of the many beliefs that held humans on a special plane, above animals and perhaps above the evolutionary process. We now know that the sensory world of animals is rich and diverse, and that for any species, unique sensory abilities have been finely tuned by evolution. There is much in the world that animals can sense but humans cannot. von Frisch led the way in developing this understanding.

SEE ALSO

Chapter 14, 1940 Orientation; Chapter 15, 1941 Bat Echolocation; Chapter 24, 1957 Psychophysical Laws.

REFERENCES AND SUGGESTED READING

Briscoe, A.D., Chittka, L., 2001. The evolution of color vision in insects. Annu. Rev. Entomol. 46, 471–510.

Dethier, V.G., 1962. To Know a Fly. Holden-Day, San Francisco.

Dyer, F.G., Arikawa, K., 2014. A hundred years of color studies in insects: With thanks to Karl von Frisch and the workers he inspired. J. Comp. Physiol. 200, 409–410.

von Frisch, K., 1914. Der Farben-und Formensinn der Bienen. Zoologische Jahrbücher (Physiologie) 35, 1–188.

Chapter 13

1938 Skinner and Learning

THE CONCEPT

Learning can best be understood by reducing distractions from the surrounding environment and using simple reward paradigms to shape behavior. Skinner's studies of learning fit with his adherence to the behaviorist school of psychological thought.

THE EXPLANATION

B. F. Skinner is best known for the Skinner box, an apparatus designed to test learning in a highly controlled environment in which cues that an animal might associate with a stimulus or task can be paired with rewards. The possible combinations and permutations of cue, task, and reward are practically infinite, and studies of learning using this paradigm have been a staple of comparative psychology for nearly a century.

Using this experimental environment, Skinner developed the concept of operant conditioning. Essentially, this is when an animal learns to operate—manipulate—its environment by responding to rewards and punishments. Skinner built this idea on the work of Ivan Pavlov, whom we credit with inventing the study simple associate learning (Pavlov's dog), and Edward Thorndike, who built the intellectual framework for understanding conditioning in the early 1900s. Skinner followed the tenets of Watson's (1913) behaviorism, which held that the inner life (contemplation, introspection) of an animal should not be considered in the study of its behavior.

Through Skinner's work we also came to understand *shaping* as a way of influencing an animal's behavior. An unsubtle approach to shaping involves the use of conditioning and rewards to train an animal to do tricks, like rolling over, playing dead, and so on. But shaping has more subtle implications in the behavior of humans and animals, suggesting that small reinforcements and punishments encountered in the course of daily life can shape an animal's, or a person's, behavior.

Important to Skinner was the intentionality of the animal's behavior. In learning through operant conditioning, the outcomes of the animal's actions

Conceptual Breakthroughs in Ethology and Animal Behavior.
DOI: http://dx.doi.org/10.1016/B978-0-12-809265-1.00013-7

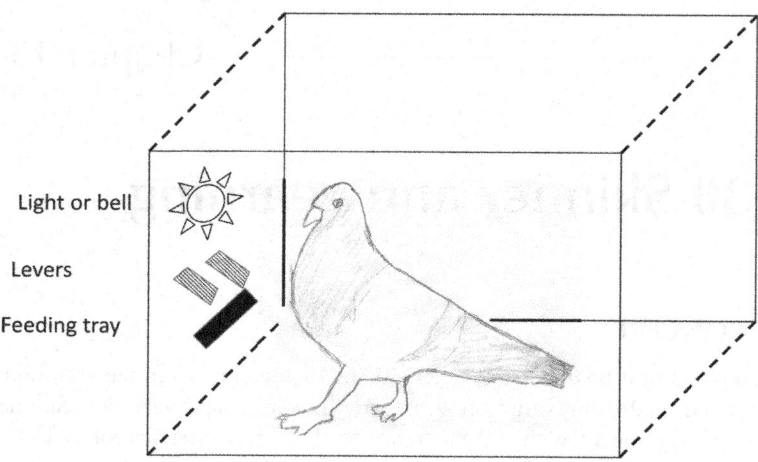

FIGURE 13.1 A pigeon in the training apparatus known as a Skinner box. There is a bell or light that can be used to build association between actions (pecking levers) and the receipt of food.

are intended. Skinner did not mean intent in the cognitive sense of forming intent, but this has been a point of confusion in the application of behavioral conditioning (Fig. 13.1).

Skinner's work formed the foundation for our modern understanding of learning and memory, including the use of conditioning in the Nobel-Prize-winning experiments of Eric Kandel on sea slugs (see Chapter 70: 1996 The Molecular Basis of Learning). However, his work also created a perception of artificiality in studies of learning performed in psychology departments in the 1950s through the 1980s that was scoffed at by ethologists and behavioral ecologists (see Chapter 10: 1882 George Romanes and the Birth of Comparative Psychology). Skinnerian approaches to the study of learning contributed to the polarity between comparative psychologists, who focused on laboratory work, and ethologists, who were carrying out field studies.

IMPACT: 9

Ethologists and behavioral ecologists have sometimes mocked the reductionist approach used in comparative psychology to study learning. From the outside this can seem to be a field that is obsessed with arguments over methodological minutiae. The Skinner box does not resemble any context in which an animal might need to learn something to survive in nature. The limited list of model species used for this work belies the word "comparative." Yet what would we know about learning without this work? The answer is not much, as field conditions do not allow for rigorous controlled experimentation needed to get at deeper mechanisms in learning and memory. In retrospect, the divide between comparative psychology and behavioral ecology in the 1960s and 1970s was unhealthy, and over the last two decades a productive integration of laboratory findings about learning over the range of behaviors seen in nature has emerged.

SEE ALSO

Chapter 10, 1882 George Romanes and the Birth of Comparative Psychology; Chapter 11, 1894 Morgan's Canon; Chapter 20, 1953 The Chasm Between Ethology and Comparative Psychology; Chapter 70, 1996 The Molecular Basis of Learning.

REFERENCES AND SUGGESTED READING

Skinner, B.F., 1938. The Behavior of Organisms. D. Appleton & Company, New York, 457 pp.

Skinner, B.F., 1951. How to teach animals. Sci. Am. 185, 26–29.

Skinner, B.F., 1953. Science and Human Behavior. Macmillan, New York, 461 pp.

Thorndike, E.L., 1905. The Elements of Psychology. A. G. Seiler, New York, 351 pp.

Watson, J.B., 1913. Psychology as the behaviorist views it. Psychol. Rev. 20, 158–177.

Chapter 14

1940 Orientation

THE CONCEPT

Animal orientation and navigation can be explained by a small number of simple rules.

THE EXPLANATION

Ethology may have reached its zenith in the early 1940s, with the focus on instinct as the prime shaper of animal behavior becoming firmly entrenched among European scientists. This period was marked by a search for commonalities of mechanisms among animals. The logic was that if instinct drives behavior, and genes underlie instinct, then the footprints of those genes will be found throughout the animal kingdom.

In line with this overall trend came the question of whether animal movements in the environment should be cataloged using a kind of movement taxonomy? Fraenkel and Gunn's (1940) book developed a systematic way of classifying animal movements. A kinesis is an undirected response to a stimulus, a taxis is oriented with respect to the location of the stimulus, and so on. More elaborately evolved animals have more complex navigational mechanisms, but all movement patterns, no matter how sophisticated, stem from the simple navigational toolkit described by Fraenkel and Gunn (1940).

The 1961 publication of a Dover edition of this book brought it back into circulation among students of animal behavior and impelled a new wave of inquiry into simple animal movements. This approach is at once satisfying—pigeonholing the behavior of diverse animals into a simple taxonomy—and frustrating, as it treats the perceptual and integrative systems that animals use in orientation and navigation as something of a black box. In this respect the approach is similar to that used in models of drive and motivation (see Chapter 25: 1960 Motivation and Drive) during the same era.

The biography of Fraenkel written by Prosser et al. (1990) is well worth reading. Fraenkel experienced all the viscissitudes of life in the 20th century. His Zionism and his scientific genius saved him from the horrors of Germany in the late 1930s and early 1940s, and ultimately an American

Conceptual Breakthroughs in Ethology and Animal Behavior.
DOI: http://dx.doi.org/10.1016/B978-0-12-809265-1.00014-9

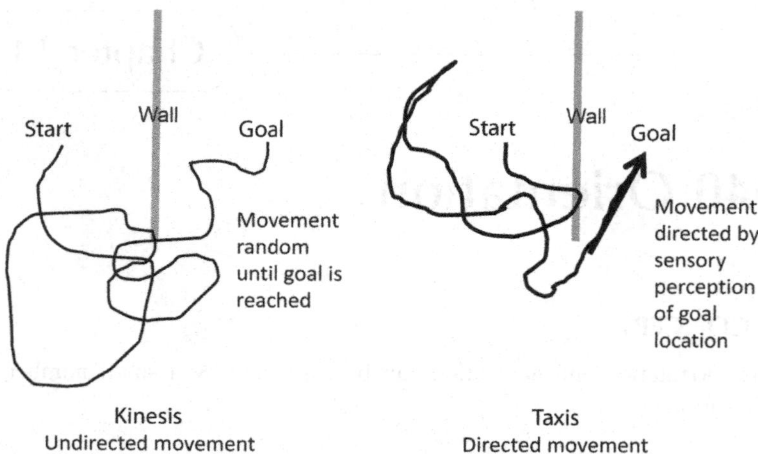

FIGURE 14.1 Animals search from a starting point to a goal, food or shelter. A wall separates the animal from its goal and eliminates the direct path between the two locations. On the left the animal lacks the ability to integrate sensory information with its movements. It uses random search—kinesis—to ultimately stumble upon its goal. On the right, the animal's movements are also initially random, as it has no sensory input from the goal. Once it can sense the goal, it integrates the sensory information with its movements to orient using a taxis.

university provided a home for his inquiring mind. As with many of the scientists highlighted in this book, Fraenkel had diverse scientific interests and his contribution to animal behavior is just a portion of the overall fabric of his scientific life.

Current studies of animal orientation and navigation are very much built on the base established by Fraenkel and Gunn, even though they are more solidly grounded in sensory physiology and neurobiology. Studies of migratory animals, such as sea turtles, monarch butterflies, and birds, currently focus on how these animals sense compass information. Once an animal can set the direction it should move in, by using a compass, then it relies on exactly the mechanisms described over 70 years ago by Fraenkel and Gunn (Fig. 14.1).

IMPACT: 5

Fraenkel and Gunn's (1940) work still shapes how animal navigation and orientation are studied. This is one of the most enduring contributions of the prewar era in animal behavior.

SEE ALSO

Chapter 12, 1914 Sensory Physiology and Behavior; Chapter 15, 1941 Bat Echolocation; Chapter 24, 1957 Psychophysical Laws.

FURTHER READING

Bell, W.J., 1991. Searching Behaviour: The Behavioural Ecology of Finding Resources. Chapman & Hall, 358 pp.

Codling, E.A., Plank, M.J., Benhamou, S., 2008. Random walk models in biology. J. R. Soc. Interface 5, 813–834.

Fraenkel, G.S., Gunn, D.L., 1940. The Orientation of Animals. Kineses, Taxes, and Compass Reactions. Clarendon Press, Oxford.

Jander, R., 1963. Insect orientation. Annu. Rev. Entomol. 8, 95–114.

Prosser, C.L., Friedman, S., Willis, J.H., 1990. Gottfried Samuel Fraenkel 1901—1984. Biographical Memoirs of the National Academy of Sciences. National Academy Press, Washington, D. C.

Chapter 15

1941 Bat Echolocation

THE CONCEPT

Bats use reflected ultrasound to navigate in unlit landscapes.

THE EXPLANATION

In the late 1700s, the Italian physiologist, Lazzaro Spallanzani discovered that bats do not use vision to navigate in the dark; his work suggested that sound played a key role in bat orientation. The exact mechanism by which bats navigate so well without light remained unknown until 1941, when Donald Griffin and Robert Galambos (Griffin and Galambos, 1941) found that bats avoid obstacles in the dark by being able to produce and receive sounds. They specifically use ultrasounds, which are sounds pitched too high for human ears to perceive. The sounds bounce back to the bats, giving them information about obstacles in their flight path; this is *echolocation*. This discovery led to a rich field of inquiry about how bats use ultrasounds for navigation, social communication, and predation.

A paper published in 1960 further revolutionized our understanding of echolocation and stimulated the growth of sensory physiology as a field (Griffin et al. 1960). Donald Griffin asked a very interesting question—do bats use their echolocation abilities to find very small prey items? He combined still and motion picture photography with sound recording to give accurate accounts of the events that occurred right before a bat captured an insect. From a 21st century perspective this does not seem too remarkable, but at the time it was quite a feat of precision technology to be able to deliver these measurements.

So what happens when a 6 g (0.2 ounce) bat pursues a 2 mg (0.000007 ounce) fruit fly? The flies are clueless about the bat's approach, as they cannot perceive the bat's sounds. At a critical distance from the fly—about 50 cm (20 inches) the bat produces a high-pitched buzz that gives it the critical echolocation information about the fly's location. After that, its all over the for fly, as the bat's and fly's paths converge.

Conceptual Breakthroughs in Ethology and Animal Behavior.
DOI: http://dx.doi.org/10.1016/B978-0-12-809265-1.00015-0

Griffin wondered if all of this was worth it for the bat. After all, a fly that size does not contain much in the way of nutrition. Based on calculations of the metabolic rate of a bat, the caloric value of a fly, and the rate at which the bats could capture flies (up to 15 per min!), Griffin realized that bats could consume about four times as much energy, by eating these very small flies, than they were using in flight. To make this work, the bat needs to be in a pretty dense swarm of flies that is not depleted too much by the bat's predatory efforts.

These discoveries led to detailed work on bats pursuing moths, as certain kinds of moth can hear bats and attempt to evade the bat by changing their flight path. In an arms race, some bat species evolved to produce sounds at different pitches than the moths can hear. In a fascinating extension of the discovery that bats can echolocate prey, Kenneth Roeder (Dethier, 1993) worked with Griffin on how moths evade bats. Other insect species produce sounds of their own, which might startle or confuse the bats (Conner and Corcoran, 2012). The evolutionary interplay between bats and their prey remains a fascinating field of study.

This discussion of bat echolocation places the focus of the development of sensory physiology as a key to understanding behavior on Donald Griffin's work, but clearly many other scientists helped to lead the way in this general field, as exemplified by Karl von Frisch's accomplishments that are highlighted in Chapter 12, 1914 Sensory Physiology and Behavior.

IMPACT: 5

As with von Frisch's studies of color vision in bees (see Chapter 12: 1914 Sensory Physiology and Behavior), Griffin's work on bat echolocation opened a world of understanding of how animals can exploit much that humans cannot even perceive. This has been a long-lasting contribution.

SEE ALSO

Chapter 12, 1914 Sensory Physiology and Behavior; Chapter 14, 1940 Orientation; Chapter 2, 1957 Psychophysical Laws.

REFERENCES AND SUGGESTED READING

Conner, W.E., Corcoran, A.J., 2012. Sound strategies: the 65-million-year-old battle between bats and insects. Annu. Rev. Entomol. 57, 21–39.

Dethier, V.G., 1993. Kenneth David Roeder 1908–1979. Biographical Memoirs of the National Academy of Sciences. <http://www.nasonline.org/publications/biographical-memoirs/memoir-pdfs/roeder-kenneth.pdf>.

Fenton, M.B., 2013. Questions, ideas and tools: lessons from bat echolocation. Anim. Behav. 85, 869–879.

Fraenkel, G.S., Gunn, D.L., 1940. The Orientation of Animals. Kineses, Taxes, and Compass Reactions. Clarendon Press, Oxford.

Galambos, R., Simmons, J.A. Echolocation in Bats. <http://www.scholarpedia.org/article/Echolocation_in_bats>.

Griffin, D.R., 1958. Listening in the Dark. Yale University Press, New Haven, CT.

Griffin, D.R., Galambos, R., 1941. The sensory basis of obstacle avoidance by flying bats. J. Exp. Zool. 86, 481–506.

Griffin, D.R., Webster, F.A., Michael, C.R., 1960. The echolocation of flying insects by bats. Anim. Behav. 8, 141–154.

Chapter 16

1947 The Evolution of Clutch Size

THE CONCEPT

The number of offspring in a clutch—the eggs laid at a given time by a bird—responds, in an evolutionary sense, to the ability of parents to care for their offspring and to constraints on population growth. Birds in populations with limited capacity to recruit new individuals evolve to reduce clutch size and to invest more in each offspring they do rear.

THE EXPLANATION

David Lack was the central figure in ornithology and the study of bird behavior in the mid-20th century. It is fair to say that his work on clutch size in birds (Lack, 1947/1948) prefigured, by two decades, the intense focus that population biologists brought to the evolution of life history characteristics in the 1960s and 1970s. Robert Macarthur, who was an avid birder, particularly followed in Lack's footsteps when he became a driving force in theoretical population biology (see Chapter 30: 1967 Island Biogeography).

Lack's study of bird clutch size (Lack, 1947/1948), which is divided into three parts, is an early and excellent example of data mining. He took advantage of an extensive existing literature on bird clutch size to shape his analysis. The information he used came from a century of careful natural history that had been painstakingly published by scholars and birders. Among other interesting findings, Lack showed that clutch size is quite consistent within species, genus and family levels of classification, but that biogeography—latitude, eastern versus western Europe, England versus Europe, islands versus mainland, temperate versus tropics—also plays a role in predicting clutch size. He was able to integrate these findings into a fascinating set of hypotheses about life history evolution. Particularly important was his conclusion that clutch sizes in the tropics are lower than in the temperate zone. This initiated a half-century of intensive scientific work on comparisons of how animals

Conceptual Breakthroughs in Ethology and Animal Behavior.
DOI: http://dx.doi.org/10.1016/B978-0-12-809265-1.00016-2

evolve in the temperate zones versus the tropics, led by such luminaries as Daniel H. Janzen.

From the last part of the 19th century through much of the 20th century, journals named for birds or ornithological societies—the *Ibis*, where Lack published this study, the *Condor*, and the *Auk*, are good examples—were outlets for massive amounts of information on the distributions, nesting habits, foraging ecology, mating behavior, and migrations of birds. Many of these journals have ceased publication or faded in prominence, but their legacy of natural history remains as a record of the world as it was during that century. Birds were not the only target of natural historians; regional entomological journals abounded, as did journals dealing with fish, reptiles, amphibia, and mammals. The loss of natural history as a focus of scientific endeavor will be sorely felt by future generations as they grapple with a changed world, and Lack's study exemplifies the kinds of gains that can be made by integrating natural history knowledge with modern evolutionary thinking.

IMPACT: 6

By studying bird clutch size, Lack (1947/1948) placed a spotlight on life history characteristics and geographical correlations with life history. His deep thinking about these topics provided an avenue for later scientists such as Cole (1954); Stearns (1976); Macarthur and Wilson (1967); and Charnov and Krebs (1973) to develop models for the evolution of life histories. He also provided a bridge from descriptive natural history to modern evolutionary biology.

SEE ALSO

Chapter 6, 1800s Birds in Their Natural Setting; Chapter 21, 1954 Life History Phenomena; Chapter 30, 1967 Island Biogeography.

REFERENCES AND SUGGESTED READING

Charnov, E.L., Krebs, J.R., 1973. On clutch size and fitness. Ibis 116, 217–219.

Cole, L.C., 1954. The population consequences of life history phenomena. Q. Rev. Biol. 29, 103–137.

Lack, D., 1947/1948. The significance of clutch size. Ibis 89, 302–352, 90,25–45.

MacArthur, R.H., Wilson, E.O., 1967. The Theory of Island Biogeography. Princeton University Press, Princeton, New Jersey, 203 pp.

Stearns, S.C., 1976. Life-history tactics-review of ideas. Q. Rev. Biol. 51, 3–47.

Chapter 17

1948 Cognitive Maps

THE CONCEPT

Some animals may navigate using an internal representation of the world—a cognitive map. They can mentally project a route to a desired location, recalculate the route if needed, and can move from spot to spot without having to retrace their steps.

THE EXPLANATION

A cognitive map is a mental representation of an animal's world that can be used to calculate optimum movement paths. This is most easily understood by considering how humans use their own cognitive map. If a person is familiar with the streets in a city, then that person can consider a starting address and an ending address and visualize a desirable route. The human can consider time and effort, traffic, and risks such as those that might be encountered when crossing intersections, in creating the mental construct of the route.

Tolman's (1948) publication set the table for testing whether nonhuman animals hold such mental representations and use them in making route decisions. Cognitive maps have become one of the holy grails of cognition. To use a cognitive map, an animal must remember detailed information about the spatial relationships of locations in their environment, be able to relate present location with desired future location, and use landmark and compass information to achieve the navigational goal. It must be able to modify the navigational plan as it is implemented to accommodate changes in the environment, such as the presence of a predator along its planned route. This is a complicated business and it is easy to see how scientists might have believed that the use of cognitive maps was restricted to humans and perhaps a few closely related species.

Studies stimulated by Tolman's paper have identified the hippocampus as the center for cognitive map information in birds and mammals (see Chapter 59: 1982 The Hippocampus and Navigation). "Place cells" within the hippocampus have particular importance in the integrative processes of

Conceptual Breakthroughs in Ethology and Animal Behavior.
DOI: http://dx.doi.org/10.1016/B978-0-12-809265-1.00017-4

cognitive mapping. Some bird and mammal species are far better at using cognitive mapping than others, depending on their ecological requirements (Wikenheiser and Redish, 2015). Some fish show three-dimensional spatial awareness that is very much like cognitive mapping (de Perera et al., 2016).

An argument has emerged over whether honeybees present a good nonvertebrate example of cognitive mapping. Honeybees are capable of solving very complicated navigational problems in ways that are analogous to the use of a cognitive map (Menzel and Greggers, 2015).

As with most behavioral capacities thought to distinguish humans from nonhumans, any distinction in cognitive mapping ability has melted away. From a behavioral perspective, the challenge is in developing experimental approaches that can conclusively test for this type of ability in animals.

IMPACT: 7

Cognitive maps are an epicenter of two important discussions. First, is cognition a unified property? For example, does employing a cognitive map rely on the same inner mechanism as social cognition? How generally are cognitive abilities distributed among species? Does the possession of those abilities by a species cause us to behave differently when interacting with that species? Cognition is one of the great research current research themes in animal behavior and Tolman's work occupies a special place in having stimulated and informed the debates.

SEE ALSO

Chapter 37, 1973 Episodic Memory; Chapter 14, 1940 Orientation; Chapter 52, 1978 Theory of Mind.

REFERENCES AND SUGGESTED READING

de Perera, T.B., Holbrook, R.I., Davis, V., 2016. The representation of three-dimensional space in fish. Front. Behav. Neurosci. 10, 40.

Menzel, R., Greggers, U., 2015. The memory structure of navigation in honeybees. J. Comp. Physiol. A Neuroethol. Sens. Neural Behav. Physiol. 201, 547–561.

Tolman, E.C., 1948. Cognitive maps in rats and men. Psychol. Rev. 55, 189–208.

Wikenheiser, A.M., Redish, A.D., 2015. Decoding the cognitive map: ensemble hippocampal sequences and decision making. Curr. Opin. Neurobiol. 32, 8–15.

Chapter 18

1948 Hormones and Behavior

THE CONCEPT

Hormones, compounds that are secreted by glands, and which can regulate behavioral expression. Hormones such as estrogen, testosterone, and insulin, have effects on development, seasonal changes, and metabolism for many body functions. Sexual development, mating behavior, bonding, and parental care are all deeply affected by hormones.

THE EXPLANATION

Humans knew that glands affect behavior thousands of years before the scientific details were worked out. Manipulation of male development and reproductive behavior through castration emerged as a technique before recorded history, and by biblical times castration of human males was a common social practice; eunuchs are mentioned in *Isaiah*, Chapter 56, verse 4, for example. Animal castration has the hallmark effect of reducing aggressive behavior and making males more manageable. Thus it remained for science to find the detailed explanation for a well-known phenomenon. By the very early 1900s it was known that extracts of testicles applied to animals changed their morphology and behavior.

With improving chemical techniques for isolating compounds from extracts of body tissues in the first half of the 20th century, a search began for the chemical structures of glandular products from the testicles, ovaries, and pancreas. In the 1920s, a team of scientists, including University of Chicago professor Frederick Koch, started a scientific pathway that led from isolating and purifying testosterone from testicles to the ultimate characterization of the structure of this compound. Adolf Butenandt and Leopold Ruzicka received the 1939 Nobel Prize in Chemistry for synthesizing testosterone. The American scientist Edward Doisy isolated and purified estrone from ovaries. Later, he shared in the 1943 Nobel Prize for Physiology and Medicine for isolating vitamin K.

The pivot point for scientific work that combined animal behavior with endocrinology came with the publication of Frank Beach's book on

Conceptual Breakthroughs in Ethology and Animal Behavior.
DOI: http://dx.doi.org/10.1016/B978-0-12-809265-1.00018-6

hormones and behavior in 1948. It is fascinating that the unscientific use of castration to render male animals that were more manageable and had tastier meat led to a key field in behavioral science. Butenandt's, Ruzicka's, and Doisy's work in isolating the active factors from ovaries and testes led to the availability of purified natural hormones and to synthetic hormones. Once the compounds were available to researchers, the door was opened to investigating how hormones regulate development and behavior within intact physiological systems.

Beach picked up on the thread of steroid hormones and behavior in the 1940s, using dogs as a study system. Because the behavioral effects of castration on dogs were well known, this system proved ideal in that it allowed him to focus immediately on understanding how the steroids extracted from testes affect behavior.

Beach's book was revelatory for behaviorists and physiologists as his synthesis led to broad considerations of how the brain, the endocrine system, and behavior interact. It also created an opportunity for using animals models such as mice, rats, and dogs as tools for studying sexual behavior, including sexual problems and reproductive health, in ways that the findings could be applied to humans.

IMPACT: 8

Beach provided a methodological framework for establishing hormonal effects on behavior. Much like Koch's postulates for pathogens, these methods provided a sound route to determining if a hormone helps to regulate a behavior.

SEE ALSO

Chapter 64, 1990 The Challenge Hypothesis.

REFERENCES AND SUGGESTED READING

Adkins-Regan, E., 2005. Hormones and Animal Social Behavior. Princeton University Press, Princeton, NJ.

Beach, F.A., 1948. Hormones and Behavior: A Survey of Interrelationships Between Endocrine Secretions and Patterns of Overt Response. Paul B. Hoeber Inc, 368 pp.

Beach, F.A., 1981. Historical origins of modern research on hormones and behavior. Horm. Behav. 15, 325–376.

Dewsbury, D., 1998. Frank Ambrose Beach April 13, 1911—June 15, 1988. National Academy of Sciences Memoirs, <http://www.nap.edu/read/9650/chapter/5>.

Ford, C.S., Beach, F.A., 1951. Patterns of Sexual Behavior. Harper, New York, 330 pp.

Lehrman, D.S., 1955. The physiological basis of parental feeding behavior in the ring dove (*Streptopelia risoria*). Behaviour 7, 241–286.

Chapter 19

1948 Information Theory

THE CONCEPT

Information can be measured and the accuracy of its transmission can be assessed. The efficiency of the passage of information through a network can also be measured.

THE EXPLANATION

What is information? Most humans view information as learned stuff, which you store in your memory. In the 1940s, the telecom industry was just coming into the modern era and Claude E. Shannon, who worked for the famous Bell Labs,[1] developed a special definition of information. He argued that information could be quantified by looking at how, in a sequence of two actions, the second action depended on the first. A perfect correspondence between the first and second acts meant that maximal information had been transmitted, while a random relationship between the two meant that no information had been transmitted. His unit of information measure was the bit.

The calculation of information transferred in an interaction is the same as entropy and is often called the Shannon–Weaver index by scientists working in ecology and behavior. In the following decades, this became one of the most-used measures of species diversity in communities, because it yields a measure that combines the number of signals (equal to species when considering diversity) and the relative frequency of responses (the number of individuals in each species).

Numerous studies also used the Shannon–Weaver index to measure the information content during bouts of communication, particularly fights

1. For readers who are too young to recognize the Bell Labs, for decades this enterprise, which started as the research arm of Bell Telephone, then was acquired by AT&T, and now is owned by Nokia, has been a center of innovation. Numerous Nobel Prizes, National Medals of Science, Turing Awards, and Kyoto Prizes have been won by Bell Lab scientists. Claude Shannon, whose work is the subject of this chapter, received the US National Medal of Science, the Kyoto Prize, the Marconi Society Lifetime Achievement Award, and is in the Inventors Hall of Fame. Bell Labs has biographical information on Shannon on its website: https://www.bell-labs.com/claude-shannon/

Conceptual Breakthroughs in Ethology and Animal Behavior.
DOI: http://dx.doi.org/10.1016/B978-0-12-809265-1.00019-8

between pairs of animals. The fact that a series of back and forth actions could be reduced to a single number was attractive, as it appeared to provide a route for comparing signaling systems across species. Unfortunately, once these numbers were calculated for a variety of species, it became apparent that little could be learned from this approach.

Shannon's definition of information does not take stored information into account. In a pairwise interaction between animals, Shannon's calculation allows us to know the information transmitted if the second animal acts upon that information. If, on the other hand, the second animal learns something for use later, then the measure is imperfect, as it does not reflect that transfer of information.

The important advance from Shannon in understanding animal communication came from seeing that immediate consequences of behavior during an interaction could be considered separately from the information stored for use in later interactions. The focus on immediate consequences ultimately led to the application of game theory to understanding strategies during animal contests (see Chapter 38: 1973 Game Theory). His work also led the way for understanding communication networks (see Chapter 76: 2002 Social Networks), an approach that has revolutionized studies of animal social systems.

IMPACT: 4

Shannon's work turned out to have major implications in research on animal behavior, mostly on the analysis of social networks, although it took many years for this impact to develop.

REFERENCES AND SUGGESTED READING

Gleick, J., 2012. The Information: A History, a Theory, a Flood. Vintage, 544 pp.

Horgan, J., 2016. Claude Shannon: Tinkerer, Prankster, and Father of Information Theory. <http://spectrum.ieee.org/computing/software/claude-shannon-tinkerer-prankster-and-father-of-information-theory>.

Shannon, C.E., 1948. A mathematical theory of communication. Bell Syst. Tech. J. 27, 379–423 and 623–656.

Chapter 20

1953 The Chasm Between Ethology and Comparative Psychology

THE CONCEPT

A deep divide developed between ethology and comparative psychology in the post–World War II era. The split played out in arguments over the relative importance of nature and nurture in behavioral development, as well as in choices of which animals to study and whether to focus on field or lab studies. The nature/nurture argument was not just a debate within the ivy-clad walls of academe, as the role of genes in determining behavior became a point of public concern. The conflict culminated in the sociobiological furor of the 1970s (see Chapter 46: 1975 Sociobiology).

THE EXPLANATION

The deep and long-lasting rift between ethology (the European school) and comparative psychology (the American school) endured from shortly after the end of World War II until the 1970s when the divide spawned the sociobiological earthquake. This was not just an argument among scientists; it was a sociological, political and philosophical *cause celèbre*. I have chosen Lehrman's (1953) paper critiquing Lorenz' ideas about innate behavior as a defining moment in this argument.

In the simplest terms, the gap between comparative psychology and ethology was the nature versus nurture argument, writ large. We now fully understand that studies of populations of animals tell us that many characteristics of living organisms (phenotypes) vary within a breeding population. Genes (nature) and environment (nurture) interact to determine the exact properties of each organism. Because each animal in a population is genetically different, and each also has at least a slightly different environment, phenotypic variation in behavior due to each factor is expected.

Conceptual Breakthroughs in Ethology and Animal Behavior.
DOI: http://dx.doi.org/10.1016/B978-0-12-809265-1.00020-4

We also now recognize that from a scientific point of view, variation among organisms—at the levels of the family, the population, and the species—is vitally interesting. We know that all variation in behavior reflects both nature and nurture. It is not one or the other, it is both, and interesting scientific inquiry delves into how much of each establishes a given behavior, and under what conditions can nurture modify the effects of nature.

The British geneticist and statistician R. A. Fisher laid the groundwork for understanding how to analyze the nature–nurture interaction, and D. S. Falconer's textbook on quantitative genetics brought these techniques to a broad audience of students. Nevertheless, Fisher became involved in the eugenics movement in the 1920s and 1930s and held views on race that would now be considered quite unacceptable in educated society (Yates and Mather, 1963).

Another factor in the debate stemmed from the desire of scientists to influence public policy. Konrad Lorenz, a leader in the study of animal behavior and one of the originators of ethology as a distinct field of inquiry had a strong penchant for self-publicity and for attempting to stretch conclusions from animal studies into recommendations for human societies. Like many Austrian men in the 1940s, he was drafted into the Wehrmacht and served the Nazi regime until he was taken prisoner by the Russians. This experience has in the minds of some placed a tinge of Nazism on Lorenz's work and made it difficult to disentangle the nature/nurture question from broader sociopolitical questions. Lorenz's role as a public intellectual invited intense scrutiny, which then fueled sometimes angry discourse about behavior.

Against the checkered sociopolitical background, the American comparative psychologist, Daniel Lehrman, whose paper is the focal point of this chapter, L. R. Aronson, T. C. Schneirla, Jay Rosenblatt, and Ethel Tobach, among others, forged an argument against assuming that all behavior is innate. Lehrman's critique of ethology (Lehrman, 1953) is about the nuts and bolts of doing science and testing hypotheses about behavior; it does not focus on the political implications of ethological theories. It is also fascinating that reports from the era surrounding Lehrman's 1953 paper indicate that Lehrman and Lorenz apparently had a cordial relationship, enjoying debating the issues.

Ironically, the early leading comparative psychologists, Romanes, Watson and Thorndike, had embraced evolution as part of their approach to analyzing behavior (see Chapter 10: 1882 George Romanes and the Birth of Comparative Psychology, and Chapter 11: 1894 Morgan's Canon). Thus embracing learned (environmental) effects as critical in shaping behavioral phenotypes represented an intellectual shift within comparative psychology. This was the behaviorism of J. B. Watson, who had come after Romanes and Thorndike in leading the field of comparative psychology.

The seed was planted during the 1950s for the opinion that ethology (and later sociobiology) disguised eugenics, racism, and biological determinism in the argument that behavior is innate. This is a debate that still rages, even though the scientific community that studies animal behavior have moved far beyond the issue. Comparative psychology has been displaced by neuroscience in most psychology departments while field studies of animal behavior have lodged in biology departments (see Chapter 10: 1892 George Romanes and the Birth of Comparative Psychology). In the meantime, as time has passed the stains on the reputations of ethology and behavioral ecology, if not on sociobiology, have been erased, in most people's minds (Fig. 20.1).

The lasting remnant of the debate is an occasional cross-disciplinary lack of communication that sometimes slows scientific progress. For an example of this, see Chapter 61, 1985 An Animal Model for Anxiety, and Chapter 63, 1990 Fear; the two chapters represent the neuroscience and behavioral ecology, respectively, of the same topic. It is evident that the developers of

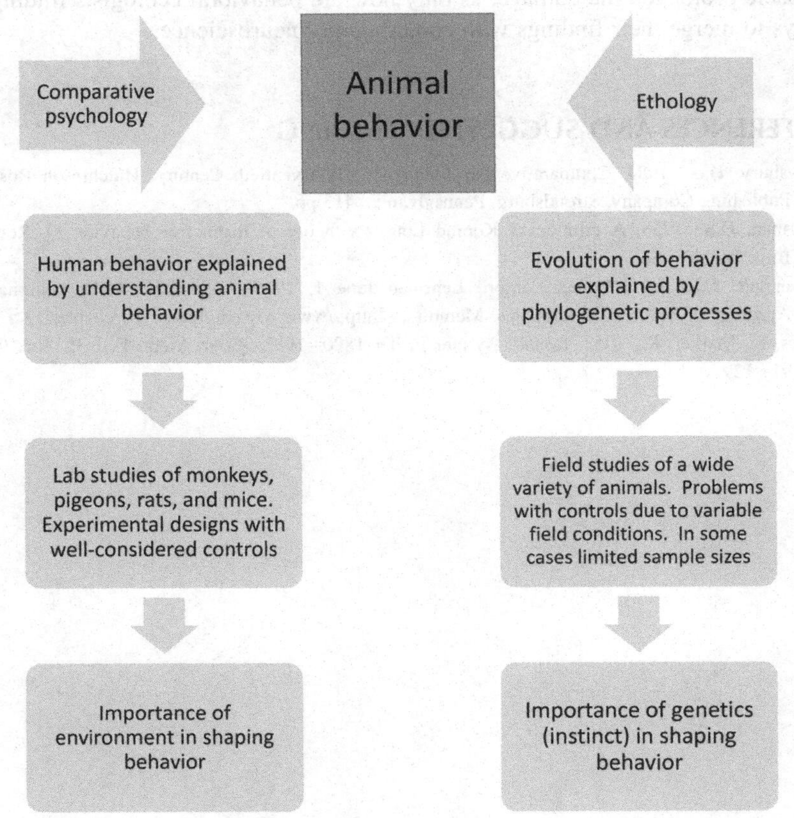

FIGURE 20.1 Major points of difference between comparative psychology and ethology, as viewed from a perspective of the time when the divide was maximal. Certainly individual investigators on both sides of the fence diverged from the stereotypes expressed in this figure.

these concepts have thought much too little about work that relates to their own findings but which seems disciplinarily distant.

In general, too little has been done to integrate related concepts in neuroscience and behavioral ecology, even when the interactions between sets of findings seem clear. How much of this myopia is carryover from decades-old disputes is not clear, but going forward it is really time to find ways to synthesize interesting findings from diverse disciplines.

IMPACT: 9

Because the debate between comparative psychologists and ethologists ultimately led to a lack of mutual respect for each others' scientific findings, this chasm harmed progress in the search of knowledge. The sociobiology debacle prolonged the damage, as only now are behavioral ecologists finding ways to merge their findings with contemporary neuroscience.

REFERENCES AND SUGGESTED READING

Dewsbury, D.A., 1984. Comparative Psychology in the Twentieth Century. Hutchinson Ross Publishing Company, Stroudsburg, Pennsylvania, 413 pp.

Lehrman, D.S., 1953. A critique of Konrad Lorenz's theory of instinctive behavior. Q. Rev. Biol. 28, 337–363.

Rosenblatt, J.S., 1995. Daniel Sanford Lehrman June 1, 1919–August 27, 1972. National Academy of Sciences Biographical Memoirs, <http://www.nap.edu/read/4961/chapter/13>.

Yates, F., Mather, K., 1963. Ronald Aylmer Fisher 1890–1962. Biogr. Mem. Fell. R. Soc. 9, 91–129.

Chapter 21

1954 Life History Phenomena

THE CONCEPT

Life history trait evolution has predictable behavioral consequences. Life history traits include the number of offspring per litter or clutch, age at first reproduction, and number of litters or clutches during a life span. Life history traits help in understanding why and how much animals invest in choosing mates. They also help to explain behavioral investments in caring for offspring. Life history properties also correlate with the degree of sociality expressed by a species.

THE EXPLANATION

I mark Lamont Cole's 1954 paper on how populations grow as the beginning of the modern ecological view of animal life histories Cole (1954). Cole opened the door to a conversation that still continues with fascinating questions like: How many young to have in a clutch? How many clutches in a year? What is the optimal reproductive life span for an animal? Studies of life history properties of animal populations answer these questions. Many analysts would focus more on Stearns' 1976 paper, which covers much of the same ground but is more often cited in the current literature (Stearns 1976). The Cole (1954) and Stearns (1976) publications bridge together and it is best to read both.

Following Malthus' lead, early ecologists used simple math to show that populations of animals grow exponentially if unchecked. An animal species increases in numbers at a rate determined by the rate of births and deaths. Ecologists call the combined effect of birth and death rate the intrinsic rate of increase of a species, or r. Of course, unchecked growth results in snowballing population size and ultimately whatever species is under discussion would cover the earth and the population would push out into space. Populations do not reach this point because of factors like disease, predation, and the capacity of the environment for supporting the population. The carrying capacity—the number of animals that the environment can support—is

Conceptual Breakthroughs in Ethology and Animal Behavior.
DOI: http://dx.doi.org/10.1016/B978-0-12-809265-1.00021-6

determined by limitations like the amount of food and number of nest sites available. Ecologists call the carrying capacity K. All of this can be expressed as algebraic equations, and plugging numbers into the equations gives us models of patterns of population growth. The models can be given more complexity by adding interactions among species and random fluctuations in the environment.

Questions about clutch size and timing of reproduction have major implications for animal behavior (see Chapter 16: 1947 The Evolution of Clutch Size) and receive considerable emphasis by Cole. The number of offspring in a clutch affects how much care can be given to each of the young. Ornithologists had long noticed that some birds have small clutches of one to two eggs while others can have clutches of eight or even more eggs. At the other extreme end is the wood duck, which lays ten or more eggs per clutch. Female wood ducks sometimes "dump" eggs into other wood ducks' nests, extending their potential clutch size by attempting to get other females to care for their young. Mammals show similar variation in their brood size.

One of Cole's major insights was that age at first reproduction has a dramatic effect on population growth rate. Species with very extended development prior to sexual maturity, like elephants, sea turtles, and humans have lower rates of population growth. In fact, age at first reproduction in human populations has as great an effect on population growth as the number of children each mother has. In many mammals, maternal age and experience have dramatic effects on offspring survival, supporting an argument that females may do best to have more than one brood in their lifetime, as later broods benefit from better, more experienced, parental care.

Cole's paper gave prominence to exploring population growth in a deep and insightful way. He built on Lack's (see Chapter 16: 1947 The Evolution of Clutch Size) studies of reproductive investment in birds. A number of studies that follow on Cole's are also highlighted in this book, including Macarthur and Wilson's exploration of r- and K-selection in their book on island biogeography (see Chapter 30: 1967 Island Biogeography). The 1970s were a particularly fertile time in population biology and Stearns' (1976) paper exemplifies how genetics can be melded with population models like Cole's.

IMPACT: 5

This is one of the most important sets of principles in behavioral ecology. Knowing how life history affects behavior is a critical goal in studies of animal behavior.

REFERENCES AND SUGGESTED READING

Blanckenhorn, W.U., 2000. The evolution of body size: what keeps organisms small? Q. Rev. Biol. 75, 385–407.

Cole, L.C., 1954. The population consequences of life history phenomena. Q. Rev. Biol. 29, 103–137. Available from: http://dx.doi.org/10.1086/400074.

Lack, D., 1947. The significance of clutch-size. Ibis 89, 302–352, 90, 25–45.

Malthus, T., 1798. An Essay on the Principle of Population. J. Johnson, London. Ebook: <http://www.esp.org/books/malthus/population/malthus.pdf>.

Martin, T.E., 1995. Avian life-history evolution in relation to nest sites, nest predation, and food. Ecol. Monogr. 65, 101–127.

Stearns, S.C., 1976. Life-history tactics: a review of ideas. Q. Rev. Biol. 51, 3–47.

van Noordwijk, A.J., de Jong, G., 1986. Acquisition and allocation of resources: their influence on variation in life history tactics. Am. Nat. 128, 137–142.

Chapter 22

1954 Zeitgebers (Time-Givers) for Biological Clocks

THE CONCEPT

External cues, or zeitgebers, keep biological clocks in synchrony with the day–night cycle. Without zeitgebers, biological clocks run slightly off an exact 24-h cycle, so over the span of a few days the internal rhythms of an animal lose their synchrony with the external world. Correct time keeping can be reestablished by restoring the zeitgebers.

THE EXPLANATION

A biological clock is self-sustaining and runs using only a small amount of energy. What it doesn't do well, on its own, is keep accurate time. Animals kept in continuous dark or light keep their rhythms for a while, but their clocks do not cycle at exactly 24 h, so without an external input the clock falls out of synchrony with the environment. Twenty-four-hour rhythms are called circadian rhythms because they cycle once a day; biological clocks typically run on a daily cycle. Animals under conditions without light input have free-running biological clocks that lose their synchrony with the environment and with other animals.

The German physiologist Jürgen Aschoff developed two key concepts that brought our understanding of biological clocks into our modern scientific framework: entrainment and the zeitgeber (Aschoff 1954). Entrainment is the synchronization of the internal clock with the external world. Clocks exist in many organs in an animal, so entrainment ensures that the clocks work well together. A zeitgeber, or time-giver, is the external cue that entrains the clock.

To understand how zeitgebers work a little background on clocks helps. Biological clocks can best be imagined as working like a mechanical clock with a pendulum or a mechanical watch with a balance wheel. Protein molecules build up and break down rhythmically and the information from this rhythm drives cycles in biological processes. In a

Conceptual Breakthroughs in Ethology and Animal Behavior.
DOI: http://dx.doi.org/10.1016/B978-0-12-809265-1.00022-8

mechanical watch the spring unloads a small bit of tension, causing the balance wheel to swing and advance a ratchet, which in turn pushes gears that lead to movement of the hands of the watch. In a biological clock small bits of energy are expended in protein synthesis to keep the molecules oscillating.

Period, or *per*, genes are found in most animals and are important in helping to drive the clock. A build-up of protein products (transcripts) within the cell from the *per* genes feeds back to turn off the *per* genes. When the *per* gene products break down the *per* genes turn back on and start production of the proteins again. Thus there is an on–off cycle for the cell.

This is where the zeitgeber comes in. Another gene, cryptochrome (*cry*) codes for proteins that are light-sensitive, particularly for blue light. *Cry* products combine with *per* products, so the whole cycle can be timed with external day–night cycles. The *cry* mechanism is somewhat different in mammals and this explanation is something of a simplification, but overall it helps to establish the mechanistic validity of Aschoff's concept of a zeitgeber (Reppert and Weaver, 2002).[1]

Another really interesting fact about biological clocks is that they are temperature independent. Many physiological processes speed up when an animal is hot and slow down when an animal is cold, but not so with circadian clocks. Given that in many habitats temperature changes with the day–night cycle, the temperature independence of the clock prevents physiological chaos. Clocks come up again in Chapter 34, 1971 Behavioral Genetics.

IMPACT: 3

Maintenance of biological clocks is a critical organismic function that ensures optimal physiological function as well as correct timing with the external world. This was one of the first mechanistic concepts developed in animal behavior and it remains one of the most important principles of behavioral physiology.

SEE ALSO

Chapter 5, 1729 Biological Clocks.

---

1. Not surprisingly, other types of light input via the eyes or the pineal gland have considerable importance in entraining circadian rhythms. Given the importance of circadian rhythms in physiology and the fact that animals have multiple clocks in their bodies, it makes sense that there is more than one way of entraining the clocks.

REFERENCES AND SUGGESTED READING

Anon. nd. The Mammalian Molecular Clock Model. HHMI BioInteractive. <http://www.hhmi.org/biointeractive/mammalian-molecular-clock-model>.

Aschoff, J., 1954. Zeitgeber der tierischen Tagesperiodik. Naturwissenschaften 41, 49–56.

Berson, D.M., Dunn, F.A., Takao, M., 2002. Phototransduction by retinal ganglion cells that set the circadian clock. Science 295, 1070–1073.

Dunlap, J.C., 1999. Molecular bases for circadian clocks. Cell 96, 271–290.

Gekakis, N., Staknis, D., Nguyen, H.B., et al., 1998. Role of the CLOCK protein in the mammalian circadian mechanism. Science 280, 1564–1569.

Herzog, E.D., Takahashi, J.S., Block, G.D., 1998. Clock controls circadian period in isolated suprachiasmatic nucleus neurons. Nat. Neurosci. 1, 708–713.

Reppert, S.M., Weaver, D.R., 2002. Coordination of circadian timing in mammals. Nature 418, 935–941.

Chapter 23

1956 The Coolidge Effect

THE CONCEPT

Sexual behavior between paired males and females diminishes over time.

THE EXPLANATION

Scientists can have humorous, if nerdy, moments. Frank Beach (see Chapter 18: 1948 Hormones and Behavior) noted that in some animal species sexual behavior between male and female pairs dwindled over time, yet if the male was presented with a new female his sexual appetite reawakened and the seemingly sexually disinterested male would actively court and copulate with the new female.

Beach named this phenomenon the "Coolidge Effect" based on a joke about President Calvin Coolidge and his wife; the story goes that both were touring a farm and that when Mrs. Coolidge observed a rooster repeatedly copulating she asked that her guide point that fact out to the president. When informed of his wife's comment, the President retorted, wondering if it was always the same hen. According to Dewsbury (1981) the first use of the phrase "Coolidge Effect" in the scientific literature was in Wilson et al. (1963), but the concept appears at least as early as Beach and Jordan's (1956) paper.

The Coolidge effect has been documented in many animal species and is linked with the dopamine reward system (see Chapter 27: 1964 Dopamine and Reward Reinforcement). A novel stimulus (the new potential sex partner) triggers dopamine, which plugs into the sexual gratification/reward system in the brain. The implications of this observation for mating systems, and for the genetic diversity of offspring are extremely important, as this mechanism provides an explanation for many animal's seemingly endless search for new sexual partners.

Conceptual Breakthroughs in Ethology and Animal Behavior.
DOI: http://dx.doi.org/10.1016/B978-0-12-809265-1.00023-X

IMPACT: 2

Amusing points like this attract the nonscientists in the general public to science. This is an excellent example of how an analogy helps to cement a novel insight about animal behavior in the public mind.

SEE ALSO

Chapter 50, 1977 The Evolution of Mating Systems.

REFERENCES AND SUGGESTED READING

Beach, F.A., Jordan, L., 1956. Sexual exhaustion and recovery in the male-rat. Q. J. Exp. Psychol. 8, 121–133.

Dewsbury, D.A., 1981. Effects of novelty on copulatory behavior: the Coolidge effect and related phenomena. Psychol. Bull. 89, 464–482.

Wilson, J.R., Kuehn, R.E., Beach, F.A., 1963. Modification in the sexual behavior of male rats produced by changing the stimulus female. J. Comp. Physiol. Psychol. 56, 636–644.

Chapter 24

1957 Psychophysical Laws

THE CONCEPT

The relationship between stimulus strength and sensory response is nonlinear. The psychophysical laws are attempts to model the relationship between sensory perception and the physical strength of a stimulus. The pressure of sound waves, for example, is measured in decibels, which are the logarithm of the energy in the sound, and sensory response matches the decibel scale better than it matches to absolute measures of sound pressure. Other sensory modalities display similar stimulus–response relationships.

THE EXPLANATION

Scientists noticed in the 19th century that the physical strength of a stimulus did not have a straight-line relationship with the strength of our perception of the stimulus. When we compare two sounds, one containing twice the energy of the other, we do not perceive the second as being twice as loud as the first. The German physiologists Ernst Weber and Gustav Fechner, working in the 1800s, developed the concept that human estimates of stimulus strength are nonlinear. Psychophysical laws allow us to assess the nonlinear nature of sensory responses to inputs like sound and light. I have pinned this principle to Stevens (1957) paper, a modern characterization of the principles first developed by Weber and Fechner.

Weber's law. Weber noted that our ability to tell whether two stimuli are different or not depends on the magnitude of the stimuli. This causes us to focus on the least notable difference between a pair of stimuli. A small difference in loudness is detectable if the initial tone was soft (low in amplitude), while that same small difference would not be noticeable if the initial tone was loud (high in amplitude).

Fechner's law. The relationship of perceived stimulus strength and actual stimulus strength is logarithmic, so that the sensory system is less able to determine differences between stimuli at high intensity than at low intensity. The decibel scale for measuring signal strength, which is often

Conceptual Breakthroughs in Ethology and Animal Behavior.
DOI: http://dx.doi.org/10.1016/B978-0-12-809265-1.00024-1

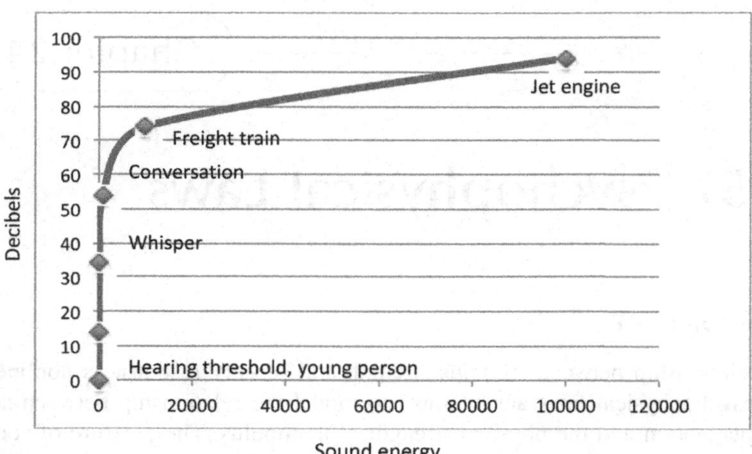

FIGURE 24.1 In this graph the actual sound energy is displayed on the *x*-axis and the perceived energy (decibels) is on the *y*-axis. At the lower end of the sound energy scale very small differences in sound energy result in large differences in the perceived strength of the sound. This allows animals to make fine discriminations among sounds within the normal range of energies found in natural environments. At high sound energies there is much less perceived difference and lower abilities to make discriminations. In other words, there are many different levels of soft sound but loud is simply loud.

applied to sound measurements, roughly corresponds to the principle of Fechner's law, although the decibel scale was developed by the Bell Labs (see Chapter 19: 1948 Information Theory) as a way of representing signal strength in telegraph and telephone transmissions. For sound measurements, each 10 units on the decibel scale represents a doubling of intensity. Natural background noise (birds calling, a babbling brook, wind rustling the leaves) is about 10 dB, while human conversation is 40–60 dB. We perceive conversation to be 3–4 times as loud as natural background noise, but in fact the sound has 1000–100,000 times more energy (Fig. 24.1).

Stevens' law. Stevens (1957) unified previous thought on psychophysical laws into a slightly more complex expression of Fechner's law:

$$\mathrm{P}=\mathrm{k}\mathrm{S}^{n}$$

where P is the perceived strength of the stimulus, k is a constant, S is the strength of the stimulus, and the exponent, n, depends on the stimulus type (e.g., sound vs brightness). There are different approaches to testing perceived strength of stimulus, the most commonly applied route is to ask subjects to compare stimulus strengths and to estimate relative magnitudes. They can also be given a pair of stimuli and be asked if the magnitudes match or not. Steven's law postulates that, as with Fechner's law, stimulus–perception relationships are nonlinear, but it accepts that the slope of the relationship differs among stimulus types.

IMPACT: 4

These laws, which seem to be intuitively true, have been the subject of considerable examination and criticism by psychologists and philosophers. Objections include the subjective methods of measuring stimulus perceptions, problems with lumping quite different sensory systems together, and the lack of consideration of cognitive processes in sensory processing. Nevertheless, psychophysical laws are quite important in understanding signal production and reception in animal behavior and they deserve a central spot in studies of the evolution of communication. The psychophysical laws have considerable explanatory power in sensory physiology and help to explain much about how animals use signals.

SEE ALSO

Chapter 12, 1914 Sensory Physiology and Behavior; Chapter 14, 1940 Orientation; Chapter 15, 1941 Bat Echolocation; Chapter 66, 1991 Receiver Psychology; Chapter 73, 1999 Multimodal Communication.

REFERENCES AND SUGGESTED READING

Anon. Weber, Ernst Heinrich, biographical entry at Encyclopedia.com. <http://www.encyclopedia.com/topic/Ernst_Heinrich_Weber.aspx#1-1G2:2830904576-full>.

Dawson, P.M., 1928. The life and work of Ernest Heinrich Weber. Phi Beta Pi Quarterly 25, 86–116.

Horeman, H.W., 1963. Neural effects and the psychophysical law. Nature 200, 1241.

Stevens, S.S., 1957. On the psychophysical law. Psychol. Rev. 64, 153–181.

Chapter 25

1960 Motivation and Drive

THE CONCEPT

Internal energy, or drive, was used to help to explain why an animal changes its activity levels. Motivation for different activities varies depending on an animal's physiological state, season, and its social environment. Motivation and drive were concepts that helped early ethologists to understand why and when an animal behaves, but they have been supplanted by modern concepts in neurobiology.

THE EXPLANATION

An animal lies on the forest floor. It stirs, shifts, and then gets up and moves purposefully, disappearing from our view. No matter whether the animal is a beetle, a snake, a fox, or a deer, it has shifted from resting to walking. The important observation is that the animal's behavioral state has changed, in this case without any perceptible change in the external environment.

Early ethologists often saw this sort of observation as a question of motivation. The animal's motivation changed, and consequently its activity changed. This reasoning led to Lorenz and Tinbergen to apply a psycho-hydraulic model of motivation in which energy for behavior pools in a metaphorical reservoir within the animal and then is drained through activity (e.g., Tinbergen, 1950).

Hinde (1959, 1960) in his papers, shifted attention to actual neurobiological mechanisms and argued against the use of motivation and drive as concepts. Much of what the theory of drive and motivation attempted to explain can be assigned to specific appetites, such as hunger, thirst, sexual desire, each of which has its own underlying regulatory pathways via hormones and neurotransmitters. Drive and motivation are black boxes that had conceptual utility when we knew little about neurobiology and the endocrine system, but which, even when Hinde was writing on this topic in 1959 and 1960, had outlived much of their utility.

Two remnants of drive and motivation theory, redirection and displacement, remain tantalizingly interesting. Redirection is when the supposed drive energy

Conceptual Breakthroughs in Ethology and Animal Behavior.
DOI: http://dx.doi.org/10.1016/B978-0-12-809265-1.00025-3

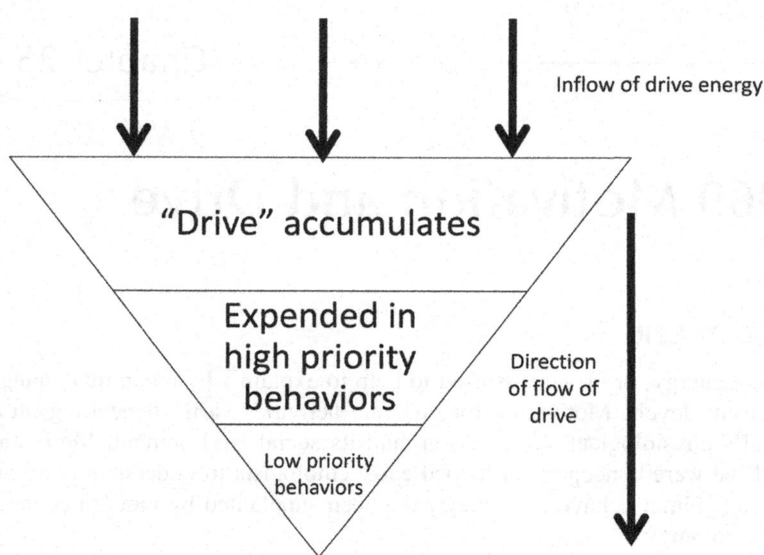

FIGURE 25.1 Drive accumulates in a reservoir and then is used as needed to fuel behavior.

is directed at a seemingly unrelated or inappropriate target. An animal in a dominance hierarchy may be subjected to aggression by the dominant in the group. Unable to retaliate, the animal attacks a lower-ranking individual, basically an innocent bystander in the initial interaction. Lashing out at inanimate objects fits into this category as well.

Displacement can refer specifically to grooming when an animal is confronted with a difficult behavior choice. This is often observed in approach/avoidance conflict, in which approaching a desirable object carries risks; an animal that pauses, attracted yet fearful, often grooms. Drive theory postulated that the grooming was an outlet for accumulated behavioral energy that could not be expended in either approach or avoidance.

Causally, displacement appears related to a broader range of self-directed behaviors (SDBs), such as pathologically excessive paw licking in dogs and feather plucking in birds. Cribbing, chewing of stall railings, by ungulates fits under this umbrella, as do repetitive behaviors, sometimes called stereotypies, such as pacing in caged animals. Neurobiologically these behaviors are related to obsessive compulsive disorders in humans; this explanation is quite a distance from the black boxes of drive and motivation (Fig. 25.1). The words drive and motivation should be used cautiously when describing animal behavior, as these reflect unknown internal devices; specific neurobiological explanations can often at least be postulated for changes in behavioral state.

IMPACT: 3

While present-day scientists should not evoke black boxes in their search for the causes of behavior, redirection and displacement remain very important behavioral expressions, and drive theory should be respected as giving science its first window into understanding these behaviors.

REFERENCES AND SUGGESTED READING

Hinde, R.A., 1959. Unitary drives. Anim. Behav. 7, 130–141.

Hinde, R.A., 1960. Energy models of motivation. Symp. Soc. Exp. Biol. 14, 199–213.

Tinbergen, N., 1950. The hierarchical organization of nervous mechanisms underlying instinctive behavior. Symp. Soc. Exp. Biol. 4, 305–312.

Chapter 26

1963 The Four Questions

THE CONCEPT

Tinbergen (1963) proposed a framework for investigating animal behavior that hinges on asking four questions: How does a behavior develop (ontogeny)? What purpose does the behavior serve (utility)? How does the behavior help the animal survive (adaptive value)? How did the behavior evolve (phylogeny)? These questions still shape research in animal behavior.

THE EXPLANATION

Science is inquiry and at its heart, inquiry is about asking questions. Let's borrow for a moment a point of view about questioning from the field of media reporting. Journalism has its interrogatives:

- Who?
- What?
- Where?
- When?
- Why?

These questions capture what the reader or viewer wants to know. Given the facts of the matter, we like to form our own opinions about events, and bad reporting is usually a failure to deliver information on the basic queries embodied in the five W's.

Animal behavior has its own set of questions that structure inquiry:

- What is the utility of the behavior and what mechanisms organize the behavior?
- How does it develop?
- How did it evolve?
- What is its adaptive value?

Niko Tinbergen, a leading animal behaviorist of the mid-20th century and one of the recipients of the 1973 Nobel Prize in Physiology or Medicine, first crystallized these questions in 1963 (Fig. 26.1).

Conceptual Breakthroughs in Ethology and Animal Behavior.
DOI: http://dx.doi.org/10.1016/B978-0-12-809265-1.00026-5

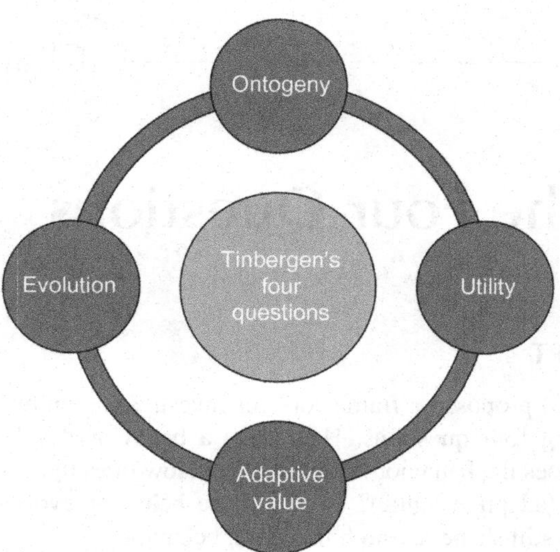

FIGURE 26.1 Asking Tinbergen's four questions about behavior is foundational to the study of animal behavior.

Oftentimes, behaviors seen in field observations are at first mysterious. As with the journalist, the scientist might ask why would an animal do that? This "why" has several dimensions. Why would a prairie dog give out a loud, high-pitched shriek when a hawk appears overhead? On the surface the behavior seems so counterintuitive, as it might well attract the hawk to the prairie dog. But delving a bit deeper, we discover that the prairie dog's call *functions* to alert other prairie dogs and that the *adaptive value* comes from enhanced survival of genetic relatives. The signal has *evolved* to be hard to locate, so less information is there for the hawk than one might think. Young prairie dogs don't give alarm calls, the inclination to give warnings *develops* as the animals mature.

The first of Tinbergen's questions, about *function and mechanism*, opens the door to thinking about whether the behavior serves that animal in foraging, mating, parenting, self-defense, or another context. Function also creates a window for thinking about mechanism—how a behavior is formed and regulated. The second, *development*, aids us in placing the behavior relative to an animal's age and sex, as well as allowing us to consider influences like seasonality, the presence of offspring, and how particular challenges might bring a behavior forward into an animal's repertoire. These two questions focus on *proximate* issues in an animal's life—the utility of the behavior on a day-to-day basis.

Evolution, on the other hand, is a process that occurs across generations in response to selective pressures. When we observe the alarm-calling prairie dog, we could ask whether related species show the same behavior. How has evolution optimized the behavior for its function? For some types of evolutionary questions we need to know the phylogeny—the family tree—of a species. For others, we need to separate environmental from genetic influences and perhaps to delve into the genes that regulate the behavior. Finally, the *adaptive value* (or survival value) of a behavior tells us whether expressing the behavior improves an animal's chances of passing the behavior on to the next generation. These two questions are about *ultimate* causes; how an animal over evolutionary time has come to be equipped with a behavior.

IMPACT: 10

The four questions are the conceptual framework for the study of animal behavior. The questions are the beginning point for all behavioral observations and experiments. As such, this is one of the most important concepts covered in this book.

REFERENCES AND SUGGESTED READING

Bateson, P., Laland, K.N., 2013. Tinbergen's four questions: an appreciation and an update. Trends Ecol. Evol. 28, 712–718.

Stamps, J., 2003. Behavioural processes affecting development: Tinbergen's fourth question comes of age. Anim. Behav. 66, 1–13.

Taborsky, M., 2014. Tribute to Tinbergen: the four problems of biology. A critical appraisal. Ethology 120, 224–227.

Tinbergen, N., 1963. On aims and methods of ethology. Zeitschrift für Tierpsychologie 20, 410–433.

Chapter 27

1964 Dopamine and Reward Reinforcement

THE CONCEPT

Dopamine is the key neurotransmitter in the reward system in the brain. Its release reinforces the repetition of behavior. Dopamine operates as a reinforcer in learning and in addiction.

THE EXPLANATION

Dopamine's role as the neural mediator of reward is so widely recognized that popular songs herald it. Yet in the 1960s, our understanding of how chemical signals in the brain—neurotransmitters—regulate behavior and mood was only in its infancy. Schultz's (1998) paper is a key link in the scientific realization that dopamine forms a central step in the reward circuit in the brain.

Knowledge of dopamine's role helps in understanding addiction, reward-seeking behavior, economic decisions, and interpersonal relationships. Dopamine does not give a sense of pleasure, but it does reinforce doing pleasurable things.

Knowledge of dopamine's role helped to open the door for investigations of other neurotransmitters, such as serotonin, gamma-amino butyric acid (GABA), acetylcholine, and glutamate. Pharmacological treatments of psychiatric disorders have become common as knowledge of neurotransmitters has grown. Also, with these findings has come the realization that neurotransmitters can have multiple roles and effects in the nervous system; thus dopamine appears also in the context of movement and movement disorders such as Parkinson's disease.

In the years since the first discovery of the roles of dopamine in reward, reinforcement and learning, the picture has become far more complicated. Scientists now recognize several dopamine receptor types. These vary in function, distribution in the nervous system, and pharmacological responses. The importance of dopamine in the reward system, though, remains unchallenged.

Conceptual Breakthroughs in Ethology and Animal Behavior.
DOI: http://dx.doi.org/10.1016/B978-0-12-809265-1.00027-7

The dopamine reward system helps us to understand how reward and punishment reinforce or extinguish behavior (see Chapter 13: 1938 Skinner and Learning). In training and shaping the behavior of companion animals, knowledge of the neuroscience of reward shows us that positive reinforcements are quite powerful in encouraging animals to display desired behaviors. Punishments, on the other hand, are handled in different ways in the brain and are not nearly as effective in shaping desired behavior. While animals can build associations between pain and specific behavioral acts, pain-based training does not engage the dopamine reward system and consequently is not nearly as effective in achieving desired behavioral results (see Chapter 61: 1985 An Animal Model for Anxiety; Chapter 63: 1990 Fear; and Chapter 65: 1991 Pain in Animals). Knowing how animals respond to reward and punishment on a neural level leads to better ways of interacting with companion animals and livestock. Understanding the dopamine reward system also helps to explain how mechanisms like trial and error learning and social reinforcement function in wild populations of animals.

IMPACT: 7

The recognition of the role of dopamine in reward and addiction was key in the development of syntheses about the neural bases for learning. The discoveries resulting from studies of the dopamine system created a model for exploration of the roles of other neurotransmitters in shaping behavior.

SEE ALSO

Chapter 13, 1938 Skinner and Learning.

REFERENCES AND SUGGESTED READING

Berridge, K.C., Robinson, T.E., 1998. What is the role of dopamine in reward: hedonic impact, reward learning, or incentive salience? Brain Res. Rev. 28, 309–369.

Schultz, W., 1998. Predictive reward signal of dopamine neurons. J. Neurophysiol. 80, 1–27.

Schultz, W., 2015. Reward. In: Toga, A.W. (Ed.), Brain Mapping: An Encyclopedic Reference, vol. 2. Academic Press, Elsevier, pp. 643–651. <https://research.pdn.cam.ac.uk/staff/schultz/pdfs%20website/2015%20Brain%20Mapping%20Encyclo.pdf>.

Stauffer, W.R., Lak, A., Kobayashi, S., Schultz, W., 2016. Components and characteristics of the dopamine reward utility signal. J. Comp. Neurol. 524, 1699–1711.

Chapter 28

1964 Inclusive Fitness and the Evolution of Altruism

THE CONCEPT

Animals can add to their fitness by aiding close relatives. Any added reproduction by those relatives that is due to the aid counts as fitness for the aid-giver. Hamilton's realization of the importance of this broader, or inclusive, view of fitness in the evolution of social behavior keyed a generation of scientists who examined genetic relationships within social groups, documented aid-giving and its outcomes, and explored the sensory abilities of animals to recognize their close kin.

THE EXPLANATION

William D. Hamilton changed scientific thought in evolutionary biology and animal behavior when he published a pair of papers on the evolution of social behavior (Hamilton, 1964). While Darwin had an inkling that aiding family members could give evolutionary benefits to both the giver and the recipient of the aid, Hamilton formalized this understanding. He captured how the possible cost of the aid, the potential benefit gained from the aid, and the genetic relatedness—kinship—between a pair of animals helps to predict whether one will aid another.

Aid-giving behavior is divided into two sorts—aid for which the donor expects a return, and aid for which the donor cannot expect a return. The first type of aid fits with Hamilton's model; there is a cost to giving the aid but also a benefit to the donor.

The second type is true altruism. True altruism is much more difficult to explain with an evolutionary model, as evolution does not favor behavior which costs an animal fitness. Incurring great personal risk when rescuing a stranger is an altruistic act for which only murky evolutionary explanations can be offered. Humans can act altruistically, but there's always room to question someone's motives when they commit an apparently altruistic act. What of stories of dolphins rescuing drowning humans? We need to

Conceptual Breakthroughs in Ethology and Animal Behavior.
DOI: http://dx.doi.org/10.1016/B978-0-12-809265-1.00028-9

understand far more about animals' minds and motivations to be able to explain such behavior.

Coming back to the issues that were more on Hamilton's mind, Hamilton explained in great detail about how genetic relatedness among animals might affect interactions and, in particular, aid giving. Hamilton also talked about the evolution of worker sterility in the eusocial insects—termites, ants, bees, and wasps. His particular observation was that because female ants, bees, and wasps derive from fertilized eggs, while males come from unfertilized eggs, evolution should favor females that help their mother to produce more females. In Hamilton's calculations, a sister has more evolutionary value than a daughter. There are important caveats to this, such as females (the queen) mating only once and there being only one queen per colony. Termites have what we would view as a more normal program of sexual reproduction in which all animals develop from fertilized eggs, so a different explanation for sterile workers is needed in that case.

Hamilton's ideas spawned an entire generation of scientific inquiry. Major lines of inquiry that leapt from Hamilton's papers include the evolutionary circumstances favoring the use of alarm signals, the ability to recognize kin and distinguish kin from non-kin, and the structure of aid-giving behavior within populations. His work also helped to spread an awareness of the importance of understanding the genetic structure of animal populations—do closely related animals tend to live near to each other or are relatives randomly distributed in the habitat?

Like all great ideas in science, there have been critics and naysayers, most prominently Edward O. Wilson (see Chapter 30: 1967 Island Biogeography) and David Sloan Wilson (see Chapter 45: 1975 Group Selection) who have argued that other evolutionary processes, particularly group selection, explain aid-giving behavior. Ultimately kin selection prevails as a powerful explanation for social phenomena.

This is the first of four chapters highlighting Hamilton's contributions to animal behavior; the others are: Chapter 36, Selfish Herds; Chapter 56, 1981 Prisoner's Dilemma; and Chapter 58, 1982 The Hamilton–Zuk Hypothesis. Hamilton was a foundational thinker whose influence in the last part of the 20th century was unparalleled. Only Darwin merits as much mention in this book. Hamilton was shy, self-effacing, and passionate about the pursuit of evolutionary explanations for natural phenomena. He died relatively young, following complications of a malarial infection he picked up while carrying out fieldwork in Africa.

IMPACT: 10

For animal behaviorists, inclusive fitness theory was the single-most-important advance in evolutionary theory since Darwin. Although the theory has generated some controversy, it has withstood both the tests of time and of intensive scrutiny.

SEE ALSO

Chapter 9, 1859 Darwin and Social Insects; Chapter 35, 1971 Reciprocal Altruism; Chapter 45, 1975 Group Selection; Chapter 46, 1975 Sociobiology; Chapter 60, 1983 Reproductive Skew.

REFERENCES AND SUGGESTED READING

Bourke, A.F.G., 2011. The validity and value of inclusive fitness theory. Proc. Biol. Sci. 278, 3313–3320.

Breed, M.D., 2014. Kin and nestmate recognition: the influence of W.D. Hamilton on 50 years of research. Anim. Behav. 92, 271–279.

Breed, M.D., Cook, C.N., McCreery, H., Rodriguez, M., 2015. Nestmate and kin recognition in honeybees. In: Aquiloni, L., Tricarico, E. (Eds.), Social Recognition in Invertebrates. Springer, pp. 147–164.

Hamilton, W.D., 1964. The genetical evolution of social behaviour I & II. J. Theor. Biol. 7, 1–52.

Hamilton, W.D., 1972. Altruism and related phenomena, mainly in the social insects. Ann. Rev. Ecol. Syst. 3, 193–232.

Liao, X., Rong, S., Queller, D.C., 2015. Relatedness, conflict, and the evolution of eusociality. PLOS Biol. <http://dx.doi.org/10.1371/journal.pbio.1002098>.

Nowak, M.A., Tarnita, C.E., Wilson, E.O., 2010. The evolution of eusociality. Nature 466, 1057–1062.

Chapter 29

1965 Harry Harlow and Social Isolation in Monkeys

THE CONCEPT

Social contact and particularly maternal care are essential for normal behavioral development in rhesus monkeys. By extension, the same level of importance of social contact in behavioral development applies in many birds and mammals.

THE EXPLANATION

How important is social contact in behavioral development? In a pioneering set of studies, Harry Harlow (Harlow et al., 1965) investigated the importance of maternal contact with young rhesus monkeys. He found that while infant monkeys would turn to dolls for comfort in the absence of other monkeys, their social development was imperfect at best, and that later in life they were unable to form normal social bonds with members of their species.

This lent support to Bowlby's *attachment theory*, which posited that normal development needs a maternal bond (Bretherton, 1992). Harlow's findings, coupled with Bowlby's theory, were used to support ideas about juvenile delinquency, criminality, and child abuse in humans. In very simple terms, under this theory animals that fail to development appropriate maternal attachments express social pathologies. Feminist scholars have criticized Bowlby's theory as placing the blame for society's ills on mothers (Vicedo, 2009).

One set of reactions to Harlow's experiments that developed much later involved discussing the humanity of the experiments. It would have been fairly obvious to the investigators that the infant monkeys were in distress during the experiment. Why would an experiment that was clearly inflicting psychological pain on young animals be continued? The answer partly lies in shifting societal attitudes toward animals and their use in research; in the 1950s and 1960s investigators held much more utilitarian attitudes toward animals, even to nonhuman primates. Also, Harlow felt that his work was extremely important and that the critical nature of the results in giving

Conceptual Breakthroughs in Ethology and Animal Behavior.
DOI: http://dx.doi.org/10.1016/B978-0-12-809265-1.00029-0

FIGURE 29.1 One of Harlow's baby monkeys, seeking support from a surrogate mother. *Harlow, H. F., 1958. The nature of love.* Am. Psychol. *13 (12), 673–685; American Psychological Association, reprinted with permission.*

insight into human behavior justified the experiment. One and a half decades into the 21st century, it is easy to find numerous scientific assessments, online blogs, and student term papers posted online that condemn this experiment, and correctly state that the investigation could not be done now. Harlow's work does not stand up well in the harsh light created by our relatively new, but deep, appreciation for the importance of respecting the cognitive lives of animals (Fig. 29.1).

IMPACT: 3

However one feels about the ethics and sociological impact of Harlow's work on social deprivation in rhesus monkeys, it inspired spirited discussion of the importance of social environment in normal behavioral development. While the experiments have been criticized as being cruel and as subjecting animals to conditions far outside the range of possibilities in nature, Harlow's work placed a strong spotlight on social factors in behavioral development, feeding into a societal discussion on the genesis of pathological and socio-pathological behavior in humans. His work left a lasting imprint on the study of behavioral development.

SEE ALSO

Chapter 10, 1882 George Romanes and the Birth of Comparative Psychology; Chapter 11, 1894 Morgan's Canon; Chapter 44, 1974 Parent–Offspring Conflict.

REFERENCES AND SUGGESTED READING

Bretherton, I., 1992. The origins of attachment theory: John Bowlby and Mary Ainsworth. Dev. Psychol. 28, 759–775.

Harlow, H.F., Dodsworth, R.O., Harlow, M.K., 1965. Total social isolation in monkeys. Proc. Natl. Acad. Sci. U S A 54, 90–97.

Suomi, S.J., van der Horst, F.C.P., LeRoy, H.A., van der Veer, R., 2008. Rigorous experiments on monkey love: an account of Harry F. Harlow's role in the history of attachment theory. Integr. Psychol. Behav. Sci. 42, 354–369.

van der Horst, F.C.P., LeRoy, H.A., van der Veer, R., 2008. "When Strangers Meet": John Bowlby and Harry Harlow on attachment behavior. Integr. Psychol. Behav. Sci. 42, 370–388.

Vicedo, M., 2009. Mothers, machines, and morals: Harry Harlow's work on primate love from lab to legend. J. Hist. Behav. Sci. 45, 193–218.

Chapter 30

1967 Island Biogeography

THE CONCEPT

The number of species on an island depends on the distance of the island from the mainland and on the size of the island. This knowledge allows scientists to predict the number of species on islands. By analogy, these predications can be made in habitats that are island-like because they are isolated from other similar habitats. Some species have life history characteristics that make them particularly successful immigrants to unpopulated islands (r-selection).

THE EXPLANATION

MacArthur and Wilson (1967) changed the worlds of animal behavior, ecology, and population biology with their monograph, *The Theory of Island Biogeography*. How did they manage to change such a broad expanse of biological thinking in one fell swoop? Part of the answer clearly lay in MacArthur's precise ability to see how biological phenomena, always messy and variable in their expression, could be captured as clear and relatively simple algebraic formulations. Another big element was Wilson's elegant writing style which, even in a scientific monograph, captured the reader and carried a rapt audience of biologists on a journey through how island size and distance from the mainland affect the number of species on an island. Along the way they introduced the key concept of r- and K-selection, an idea that buoyed population biology through the 1970s (see Chapter 21: 1954 Life History Phenomena).

MacArthur and Wilson (1967) addressed core questions in ecology. These included how many species a habitat could support, a problem they approached by considering the relationship between the number of species on islands and those islands' areas. Species-area curves remain a fundamental concept. They also analyzed how distance from the mainland affects the equilibrium number of species on an island. Perhaps most importantly, they considered the life history characteristics of species that were successful colonizers on islands. This line of reasoning supported the concept of r- and K-selection, in which r-selected species have life history traits that favor

Conceptual Breakthroughs in Ethology and Animal Behavior.
DOI: http://dx.doi.org/10.1016/B978-0-12-809265-1.00030-7

dispersal, invasion, and the potential for explosive population growth. K-selected species have life history traits that favor ability to compete in habitats that are already occupied and perhaps crowded.

MacArthur was a mathematician-turned-biologist and carried a birder's love of the mysteries of bird species distributions into his orderly world of using equations to describe physical phenomena. If any one person created and defined the area of population biology, it was MacArthur, who also produced important works on optimal foraging and niche partitioning. He died young, in 1972, and left a lasting imprint on ecology and animal behavior.

E. O. Wilson's abilities to write and synthesize massive topics have carried forward in a long career and the publication of numerous other books, including *The Insect Societies* (1971), *Sociobiology: The New Synthesis* (1975) and, with Bert Holldbobler, *The Ants* (1990). *The Ants* was awarded a Pulitzer prize in 1991. Wilson has become a leading public intellectual as well as being a scientific figure. His interest in engaging larger topics such as the philosophy of human interactions with nature, the biology of belief and religion, and the controversy over whether human behavior is more governed by genetics (evolution) or learning has brought him both praise and criticism. (See Chapter 46: 1975 Sociobiology, for more on Wilson and the public controversy about nature vs nurture in shaping human behavior.)

IMPACT: 6

That two of the scientific geniuses of 20th century came together to produce this paradigm-shaping book was good fortune for generations of animal behaviorists and ecologists. The publication of *The Theory of Island Biogeography* was a watershed moment in the history of biology.

SEE ALSO

Chapter 16, 1947 The Evolution of Clutch Size; Chapter 21, 1954 Life History Phenomena; Chapter 44, 1974 Parent–Offspring Conflict; Chapter 53, 1980 Dispersal.

REFERENCES AND SUGGESTED READING

Chave, J., Muller-Landau, H.C., Levin, S.A., 2002. Comparing classical community models: theoretical consequences for patterns of diversity. Am. Nat. 159, 1–23.

Gotelli, N.J., 2000. Null model analysis of species co-occurrence patterns. Ecology 81, 2606–2621.

Gustafson, E.J., Gardner, R.H., 1996. The effect of landscape heterogeneity on the probability of patch colonization. Ecology 77, 94–107.

Losos, J.B., Schluter, D., 2000. Analysis of an evolutionary species–area relationship. Nature 408, 847–850.

MacArthur, R.H., Wilson, E.O., 1967. The Theory of Island Biogeography. Princeton University Press, Princeton, NJ.

Whittaker, R.J., Willis, K.J., Field, R., 2001. Scale and species richness: towards a general, hierarchical theory of species diversity. J. Biogeogr. 28, 453–470.

Chapter 31

1968 Tool Use

THE CONCEPT

Some nonhuman animals modify objects in their environment so they can serve as levers, probes, pincers, or other types of tools. The development of this concept broke down an artificial division that set humans apart from other animals.

THE EXPLANATION

Humans have often wished to find some simple set of properties that separate us from all other animals. The roots of this desire are complex and include: (1) an impulse to be better than other creatures; (2) religious belief that man was created separately from animals; and (3) moral justifications for consuming, domesticating, and sometimes mistreating other species. From a biological point of view, humans are animals and looking for evidence that humans are somehow not animals is an obvious exercise in futility.

The evidence to separate humans from animals has, over the years, been given as bipedal stance, large brain relative to body size, tool use, cognition, and morality. We now know that none of these characteristics are uniquely human. In fact, the hardest human behavioral motivation to find in animals seems to be spite, not exactly a feature of which we should be proud. But even spite shows up in some monkeys if we prompt them in the right way (Hemelrijk and Puga-Gonzalez, 2012; Leimgruber et al., 2015).

When Jane Goodall started her pioneering studies of chimpanzees in the early 1960s, much was still made of the boundaries between humans and other animals (van-Lawick-Goodall 1968). As she made detailed behavioral observations of chimpanzees it became clear that humans have plenty in common with their closest evolutionary relative, and that a lot about this commonality erodes the view of human distinctiveness.

Reports of her studies trickled out through the mid-1960s and her work, as well as her personality, captured the public imagination. A most fascinating finding was that chimpanzees do, in fact, use tools to accomplish important tasks in their lives. They modify twigs and leaves for use in hunting termites, and they use sticks and rocks as levers and hammers (van-Lawick-Goodall

Conceptual Breakthroughs in Ethology and Animal Behavior.
DOI: http://dx.doi.org/10.1016/B978-0-12-809265-1.00031-9

1968). Her work played a key role in breaking the public's sensibility, held strongly in much of the world in the 1950s, about the supposed special position of humans in the natural world.

Goodall gathered her observations into a book-length scientific report published in 1968.[1] Since Goodall's (van-Lawick-Goodall 1968) publication, other animals have been shown to use tools as well. Most notable are New Caledonian crows, which have remarkable abilities to fashion tools and use them in novel ways (Weir et al., 2002). Goodall's publication was groundbreaking in many ways. It showed that determined fieldwork could yield breakthrough findings about our primate relatives. It asserted the value of identifying animals in a population individually and following their behavior from year to year as those animals grew and matured. And, it established the validity of habituating wild animals to the presence of an observer for the purpose of collecting behavioral data.

IMPACT: 4

Goodall remains one of the best-known and most revered figures in animal behavior. Her focus has turned to primate conservation, community engagement in conservation through her Roots and Shoots program, and in bringing her messages to the broader public through books and lectures. The acceptance that animals can use tools was one of the first steps in crediting nonhuman species with complex cognitive lives.

SEE ALSO

Chapter 11, 1894 Morgan's Canon.

REFERENCES AND SUGGESTED READING

Hemelrijk, C.K., Puga-Gonzalez, I., 2012. An individual-oriented model on the emergence of support in fights, its reciprocation and exchange. PLOS One 7, e37271.

Hunt, G.R., 1996. Manufacture and use of hook-tools by New Caledonian crows. Nature 379, 249–251.

Leimgruber, K.L., Rosati, A.G., Santos, L.R., 2015. Capuchin monkeys punish those who have more. Evol. Hum. Behav. Available from: http://dx.doi.org/10.1016/j.evolhumbehav.2015.12.002.

Taylor, A.H., Hunt, G.R., Holzhaider, J.C., Gray, R.D., 2007. Spontaneous metatool use by New Caledonian crows. Curr. Biol. 17, 1504–1507.

Van Lawick-Goodall, J., 1968. The behaviour of free-living chimpanzees in the Gombe stream reserve. Anim. Behav. Monogr. 1, 161–311.

Weir, A.A.S., Chappell, J., Kacelnik, A., 2002. Shaping of hooks in New Caledonian crows. Science 297, 981.

1. The citation technically is Van Lawick-Goodall (1968) because she was using her husband's name at the time.

Chapter 32

1969 Territoriality and Habitat Choice

THE CONCEPT

Animals divide space and other resources into territories. In the simplest definition, a territory is any defended space (Klopfer, 1969). Understanding territoriality gives great insight into mating systems and patterns of association among animals in social groups.

THE EXPLANATION

For animals, territories are nearly always coupled with resources. Territories also often determine mating opportunities and, in the larger picture, drive the distribution of animals within a habitat. Territoriality in animals raises many interesting questions, such as whether combat is universal in animals, including humans; what determines the winner of territorial conflict; and whether conflict behavior has evolved to allow animals to avoid injury (ritualized combat).

Klopfer's (1969) book was seminal in giving scientists working on animal behavior access to information about how territories were established and held by animals. It helped to stimulate a decade or more of intensive study of territoriality and aggression. At the same time, Fretwell and Lucas (1969) provided a mathematically detailed theoretical framework for understanding the ecology of territoriality.

The interest of animal behaviorists in the 1960s in territoriality and aggression interacted with a concurrent major public conversation. The 1960s were fraught with public discussion about human aggression, the inevitability of war, and the roles of nature versus nurture in shaping conflict among humans. Serious biologists, like the Nobel-Prize-winning Austrian ethologist Konrad Lorenz, contributed to the public conversation with books (e.g., Lorenz, 1963) geared to audiences of nonscientists. Nonscientists, including Ardrey (1966) and Morris (1967) wrote passionately on the topic

Conceptual Breakthroughs in Ethology and Animal Behavior.
DOI: http://dx.doi.org/10.1016/B978-0-12-809265-1.00032-0

of animal aggression, human aggression, and the prospects for peaceful existence among humans.

Animal behaviorists embraced this discussion, sometimes with serious studies of behavior in animals that allowed comparisons across habitats and among species (e.g., Fretwell and Lucas, 1969). An entire scientific journal, *Aggressive Behavior*, is devoted to the comparative psychology and neuroscience of conflict among animals and humans. The public debate over the inevitability of conflict among humans was re-ignited by chapters in Wilson's (1975) book *Sociobiology*; the chronicle of that conversation continues in Chapter 46, 1975 Sociobiology. Territoriality is a complex issue, if for no other reason than the fact that so many different types of animals are territorial. While territorial aggression is only one context in which animals may express aggressive behavior, understanding territoriality remains key to the study of conflicts among animals.

IMPACT: 3

Territoriality was well known and the impact of the Fretwell and Lucas (1969) paper coupled with the Klopfer (1969) book was to bring a strong scientific response to the popular discussion of human aggression stimulated by Konrad Lorenz's book, *On Aggression* (1963), and Robert Ardrey's book *The Territorial Imperative* (1966). These scientific works fed into work on animal conflict by Maynard Smith (see Chapter 41: 1973 Animal Conflict) and also help to inform thinking about mating systems (see Chapter 50: 1977 The Evolution of Mating Systems).

SEE ALSO

Chapter 21, 1954 Life History Phenomena; Chapter 44, 1974 Parent–Offspring Conflict; Chapter 50, 1977 The Evolution of Mating Systems; Chapter 53, 1980 Dispersal.

REFERENCES AND SUGGESTED READING

Ardrey, R., 1966. The Territorial Imperative. Atheneum, 390 pp.

Fretwell, S.D., Lucas, H.L., 1969. On territorial behavior and other factors influencing habitat distribution in birds. Part 1 theoretical development. Acta Biotheor. 19, 16–36.

Klopfer, P.H., 1969. Habitats and Territories: A Study of the Use of Space by Animals. Basic Books.

Lorenz, K., 1963. On Aggression. Harcourt Brace & Company, New York.

Morris, D., 1967. The Naked Ape. Jonathan Cape, London, 252 pp.

Wilson, E.O., 1975. Sociobiology: The New Synthesis. Harvard University Press.

Chapter 33

1970 Sperm Competition

THE CONCEPT

When sperm from more than one male are present, competition occurs and selection favors production of sperm with strong competitive abilities. Sperm competition also favors male phenotypes for larger ejaculate volumes and higher sperm counts in semen.

THE EXPLANATION

Evolutionary success does not stop with finding a mate, courting, and copulating. In many animals sperm from multiple males can be present, and the competition for producing young extends to the swimming ability of the sperm, in which speed and endurance play important roles. Parker's (1970) review synthesized what was known, at that point, about sperm competition in insects, but more importantly, he laid out general principles and hypotheses that impacted research on sperm competition over the following decades.

Many more primitive animals (technically, those in more basal clades) practice external fertilization. Males release their sperm in water and the sperm then swim to eggs that females have dispersed. In some cases the male and female never come into proximity; they just release their gametes into the surroundings and hope for the best. In other cases, including many species of fish, courtship precedes sperm and egg release so that the gametes are close to each other and some degree of mate choice occurs. While more rare in terrestrial animals, external fertilization is the rule in frogs, and courtship plays an important role prior to gamete release. External fertilization opens the door for sperm competition and selection favors males that produce fast and enduring sperm.

In most terrestrial animals internal fertilization occurs. Males have an intromittent organ for copulation and sperm are deposited directly in the female's reproductive tract. This can be viewed as an adaptation for life on land, as sperm deposited in a terrestrial environment would be hapless. Internal fertilization also helps to ensure paternity for a given male, and the

Conceptual Breakthroughs in Ethology and Animal Behavior.
DOI: http://dx.doi.org/10.1016/B978-0-12-809265-1.00033-2

male can augment paternity certainty by guarding the female from the completion of copulation until fertilization is certain. In many species, though, mechanisms like mate guarding and copulatory plugs fail or the male simply departs after copulation, and females copulate with more than one male, resulting in sperm competition within their reproductive tracts. As with external fertilization, speed and endurance become key, and natural selection has acted to make sperm effective competitors. In species in which sperm competition is likely, i.e., females mate with more than one male, production of more sperm and larger ejaculate volumes are also important outcomes of sperm competition.

IMPACT: 5

By placing the focus on what happens between copulation and fertilization, Parker opened a new and exciting dimension for understanding sexual selection and evolution.

SEE ALSO

Chapter 50, 1977 The Evolution of Mating Systems.

REFERENCES AND SUGGESTED READING

Arnqvist, G., Nilsson, T., 2000. The evolution of polyandry: multiple mating and female fitness in insects. Anim. Behav. 60, 145–164.

Chapman, T., Arnqvist, G., Bangham, J., Rowe, L., 2003. Sexual conflict. Trends Ecol. Evolut. 18, 41–47.

Jennions, M.D., Petrie, M., 2000. Why do females mate multiply? A review of the genetic benefits. Biol. Rev. 75, 21–64.

Parker, G.A., 1970. Sperm competition and its evolutionary consequences in insects. Biol. Rev. Camb. Phil. Soc. 45, 525–567.

Parker, G.A., Pizzari, T., 2010. Sperm competition and ejaculate economics. Biol. Rev. 85, 897–934.

Simmons, L.W., 2001. Sperm Competition and Its Evolutionary Consequences in the Insects. Princeton University Press, 456 pp.

Smith, R.L., 1984. Sperm Competition and the Evolution of Animal Mating Systems. Academic Press.

Wedell, N., Gage, M.J.G., Parker, G.A., 2002. Sperm competition, male prudence and sperm-limited females. Trends Ecol. Evol. 17, 313–320.

Chapter 34

1971 Behavioral Genetics

THE CONCEPT

Behavioral phenotypes are often correlated with specific genotypes. Modifications of genes can affect behavior and the study of mutations has given great insight into how the expression of behavior is regulated. Artificial selection, the process that leads to domestication, is also a powerful tool in behavioral genetics. Molecular genetics has added many techniques for exploring how behavioral phenotype reflects an animal's underlying genetics.

THE EXPLANATION

Behavioral genetics, as a field, is as old as the human realization that the behavior of domesticated animals could be manipulated through controlled breeding in the same way as traits such as stature, muscularity, and coat color (see Chapter 2: 12,000 Years Before Present, Domestication). Selecting dogs for behavioral traits dates back to at least the times of the Egyptian pharaohs. Darwin (see Chapter 8: 1859 Darwin and Behavior) provided a natural selection context for understanding the effects of artificial selection by humans on animal traits.

The great *Drosophila* geneticists of the 20th century—T. H. Morgan, H. J. Muller, and T. G. Dobzhansky—all used behavioral phenotypes within their arsenal of tools for studying genetics. Morgan was awarded the 1933 Nobel Prize in Physiology or Medicine for his fundamental contribution to our understanding of chromosomes as the vehicles that carry genes. Muller received the 1946 Nobel Prize in Physiology or Medicine for discovering that X-rays could cause mutations. Dobzhansky contributed in elegant ways to the "modern synthesis" in evolutionary biology. Each added much, through their work, to what we know about how genes influence behavior.

General knowledge of behavioral genetics was assembled in the Scott and Fuller (1965) book on behavioral genetics of dogs. In the 1980s, Robert Plomin and his colleagues published a textbook on behavioral genetics, which has gone through multiple editions (Plomin et al., 1989). This book

Conceptual Breakthroughs in Ethology and Animal Behavior.
DOI: http://dx.doi.org/10.1016/B978-0-12-809265-1.00034-4

increased acceptance of behavioral genetics as part of the curriculum in many psychology departments.

The critical conceptual turning point for this field came with the publications of Konopka and Benzer's (1971) paper on clock mutants in *Drosophila*. In a very prescient manner, Konopka and Benzer provided the essential link between the classical behavioral genetics of Morgan and Muller, which focused on phenotypes and chromosomes, and the potentialities of using phenotype to expose the molecular workings of a behavioral trait. Little was known in 1971 about the regulatory genome and nothing was available on genetic sequences. Yet Konopka and Benzer (1971) showed how these tools would come into play over the next 40 years in linking behavioral phenotypes to genetic mechanisms. Konopka and Benzer found that changes in the same gene could affect the fly's clock in different ways, a finding that sparked a search for specific mutation effects on clock gene-related behavior and ultimately for the molecular mechanisms underlying this behavior.

IMPACT: 9

Behavioral genetics links the deep past of humans, in which the use of artificial selection allowed domestication of animals, with modern biology, in which tools from molecular genetics inform how we apply Tinbergen's four questions—phylogeny, utility, adaptive value, and development—to a myriad of topics in animal behavior.

SEE ALSO

Chapter 2, 12,000 Years Before Present Domestication; Chapter 8, 1859 Darwin and Behavior; Chapter 34, 1971 Behavioral Genetics; Chapter 38, 1973 Game Theory.

REFERENCES AND SUGGESTED READING

Benzer, S., 1971. From the gene to behavior. JAMA 218, 1015–1022.

Harris, W.A., 2008. Seymour Benzer 1921–2007. The man who took us from genes to behaviour. PLOS. <http://dx.doi.org/10.1371/journal.pbio.0060041>.

Konopka, R.J., Benzer, S., 1971. Clock mutants of *Drosophila melanogaster*. Proc. Natl. Acad. Sci. USA 68, 2112–2116.

Plomin, R., DeFries, J.C., McClearn, G.E., 1989. Behavioral Genetics: A Primer. WH Freeman & Co.

Scott, J., Fuller, J., 1965. Genetics and the Social Behavior of the Dog. University of Chicago Press, Chicago.

Chapter 35

1971 Reciprocal Altruism

THE CONCEPT

Behavioral contracts between animals can establish "I'll help you now if you'll help me later" arrangements. In the absence of verbal negotiations, evolution can shape behavior that conforms to this kind of agreement.

THE EXPLANATION

Trivers (1971) developed the idea that animals might enter into contracts, so that aid given by one animal to another would be reciprocated later in time; this is called reciprocal altruism. Reciprocation could require cognitive awareness of the deal being made, which is probably a large leap for many animals and also difficult to measure. Alternatively, evolution could drive mechanisms by which reciprocation is built into social interactions. In order to be an effective social mechanism, reciprocation must involve some sort of enforcement or penalty, such as social ostracization or physical punishments for animals that fail to reciprocate.

This paper is part of a flurry of interest sparked by Hamilton's 1964 work on the evolution of altruism (see Chapter 28: 1964 Inclusive Fitness and the Evolution of Altruism). While a 17-year gap seems quite lengthy, Hamilton's thinking was so far ahead of his time that it took a nearly a decade for theoretical and empirical responses to develop. In addition to this paper, which added a critical idea to the mix in the discussion about the evolution of altruism, seminal contributions around the same timeframe were made by Lin and Michener (1972), Alexander (1974), Wilson (1975), and West-Eberhard (1975). It was difficult for me, among the papers by Trivers, Lin and Michener, Alexander, and West-Eberhard to choose which one to highlight, but Trivers' work stands out for its novelty, as well as for being published first. The paper by D. S. Wilson on group selection is discussed in Chapter 45, 1975 Group Selection.

Cheating is a major fly in the ointment in thinking about reciprocal altruism. If an individual has behaved to benefit another animal, what prevents that animal from ignoring the donor's future needs? Complex social structures

Conceptual Breakthroughs in Ethology and Animal Behavior.
DOI: http://dx.doi.org/10.1016/B978-0-12-809265-1.00035-6

in which animals have a cognitive grasp of their social histories lend themselves to enforcement of reciprocation. An animal that fails to reciprocate may be shunned or subjected to physical punishment. This indeed appears to be the case in at least some primates (Clutton-Brock and Parker 1995).

Trivers' ideas flow directly into the later application of the prisoner's dilemma to thinking about conditions under which animals might cooperate with one another (see also Chapter 56: 1981 Prisoner's Dilemma).

IMPACT: 3

This idea was part of the avalanche of ideas about how altruism might evolve and it played a key role in debates that continued for decades. Empirical examples of reciprocal altruism are scarce, perhaps absent altogether, but this does not diminish the importance of the concept in shaping the debate about the evolution of altruism.

SEE ALSO

Chapter 28, 1964 Inclusive Fitness and the Evolution of Altruism; Chapter 36, Selfish Herds; Chapter 41, 1973 Animal Conflict; Chapter 45, 1975 Group Selection; Chapter 56, 1981 Prisoner's Dilemma.

REFERENCES AND SUGGESTED READING

Alexander, R.D., 1974. The evolution of social behavior. Annu. Rev. Ecol. Syst. 5, 325–383.

Axelrod, R., 1984. The Evolution of Cooperation. Basic Books (19 January 1995); <http://dx.doi.org/10.1038/373209a0>.

Axelrod, R., Hamilton, W.D., 1981. The evolution of cooperation. Science 211, 1390–1396.

Clutton-Brock, T.H., Parker, G.A., 1995. Punishment in animal societies. Nature 373, 209–216.

Lin, N., Michener, C.D., 1972. Evolution of sociality in insects. Q. Rev. Biol. 47, 131–159.

Trivers, R.L., 1971. The evolution of reciprocal altruism. Q. Rev. Biol. 46, 35–57.

West-Eberhard, M.J., 1975. The evolution of social behavior by kin selection. Q. Rev. Biol. 50, 1–33.

Wilson, D.S., 1975. A theory of group selection. Proc. Natl. Acad. Sci. USA 72, 143–146.

Chapter 36

1971 Selfish Herds

THE CONCEPT

Animals in groups should compete for positions that minimize their exposure to risk. Generally speaking, central positions are preferred as other group members then serve as a buffer against threats.

THE EXPLANATION

Lots of animal species group together in herds, flocks, schools, aggregations, and the like. Until the 1970s, most observers would have said that grouping gives advantages in having multiple eyes watching for predators, but since then we have learned a lot more about groups and why animals might want to be close to each other. In an altruistic world, each animal contributes to the group's safety, and animals share in the benefits from the increased protection.

When you closely observe a herd, fish, or school, it turns out that oftentimes the benefit isn't as equally shared as might first be guessed, because position in the group has a major effect on risk level. There are good spots in the group, often in the middle, and bad spots in the group, often at the sides and trailing edges. Thus even though the many eyes argument makes sense (see Chapter 39: 1973 Many Eyes Hypothesis), we would also expect a lot of jostling and maneuvering to win the best spots within the crowd. Selfishly, an animal may prefer to be with a group only if it has a good chance at being in a favored position. Much of what we now understand about groups centers on this kind of selfish motivation for being together, an insight first promoted by Hamilton's (1971) paper.

It is important to consider here that sometimes we see a flock or herd, such as Canada geese or wild horses, in which individuals at the edge of the group are most watchful or protective. In Canada geese, it appears that there is a certain amount of turn-taking among individuals in a flock, so perhaps the multiple eyes argument can apply, but perhaps animals in subgroups in the flock are family members, so the watchful geese are aiding relatives, rather than random birds. In wild horses, the protective animal is the herd's stallion and while he may take on mountain lions or other predators,

Conceptual Breakthroughs in Ethology and Animal Behavior.
DOI: http://dx.doi.org/10.1016/B978-0-12-809265-1.00036-8

FIGURE 36.1 Whenever a group of animals is observed together, the question of whether the group is structured selfishly, with each animal jockeying for advantageous position, should be considered. *Photo: Courtesy of Madison Sankovitz.*

his main motivation is to prevent other males from interloping into his harem. Selfishness abounds, but sometimes the reasons for being selfish stretch beyond geometry.

A huge value in grouping is that when under threat, the animals can create visual confusion by scattering. It takes a very focused predator to settle on one potential prey and stick with the chase after that prey when dozens or hundreds of other potential prey are zooming at random directions with the predator's visual field. Taking advantage of this is not precisely selfish, but in a scattering flock, each animal is acting for its own benefit, not for its group mates. In fact, it is taking advantage of the fact that risk is distributed to other animals.

A more recent spin on understanding selfishness in herds and flocks stems from the concept of public information (see Chapter 79: Public and Private Information). When an animal does something that can be seen, heard, or smelled by other animals, it has created public information. Public information is exploitable if it gives clues to the location of offspring, food, or other valuable resources. Why bother to forage when you can discover food locations by watching other animals forage? Gulls and crows are particularly good examples of birds that seem to flock together in order to take advantage of public information created by their flock-mates. Selfish herds remain a rich avenue for investigation (Fig. 36.1).

IMPACT: 7

The lovely thing about this concept is the intuitive sense behind it. It is easy to understand why animals jockey for position when in groups, how being in a group may involve a balance between risk and protection, and how differences in exposure to risk drive behavior.

SEE ALSO

Chapter 28, 1964 Inclusive Fitness and the Evolution of Altruism; Chapter 35, 1971 Reciprocal Altruism; Chapter 39, 1973 Many Eyes Hypothesis; Chapter 45, 1975 Group Selection; Chapter 79, Public and Private Information.

REFERENCES AND SUGGESTED READING

Danchin, E., Giraldeau, L.A., Valone, T.J., Wagner, R.H., 2004. Public information: from nosy neighbors to cultural evolution. Science 305, 487–491.

Foster, W.A., Treherne, J.E., 1981. Evidence for the dilution effect in the selfish herd from fish predation on a marine insect. Nature 293, 466–467.

Gueron, S., Levin, S.A., Rubenstein, D.I., 1996. The dynamics of herds: from individuals to aggregations. J. Theor. Biol. 182, 85–98.

Hamilton, W.D., 1971. Geometry for the selfish herd. J. Theor. Biol. 31, 295–311.

Parrish, J.K., Edelstein-Keshet, L., 1999. Complexity, pattern, and evolutionary trade-offs in animal aggregation. Science 284, 99–101.

Roberts, G., 1996. Why individual vigilance declines as group size increases. Anim. Behav. 51, 1077–1086.

Ward, P., Zahavi, A., 1973. Importance of certain assemblages of birds as information-centers for food-finding. Ibis 115, 517–534.

Chapter 37

1973 Episodic Memory

THE CONCEPT

The brain can represent past events as episodes in which visual, auditory, olfactory, and other associations with a situation can be recalled and used to project potential future events.

THE EXPLANATION

Humans know from self-assessments of our own mental processes that we can reflect on past events and, based on previous occurrences, make plans for the future (Tulving and Thomson, 1973). This has been termed mental time travel; the ability to imagine the future as informed by what took place in the past. Mental time travel relies on episodic memory, which captures events and their contexts.

For humans such planning is critical to how we conduct our lives. We might imagine, in the morning, the coming events in our day, organize the order of events to be efficient, and imagine how we will perform actions, engage in conversations, and travel during the day. These imaginings rely on episodic memory.

The terms *chronesthesia* and *mental time travel*, the conscious awareness of the subjective nature of the passage of time, have also been applied to the use of episodic memory in planning. These are viewed as higher-order cognitive processes, along with theory of mind, which is discussed in another chapter (see Chapter 52: 1978 Theory of Mind).

Humans can think introspectively about the use of mental time travel and describe our thought processes. As with theory of mind, cognitive thinking on this level is very difficult to assess in nonhuman animals, as scientists must look to the behavior of an animal in order to deduce the thought processes used to shape that behavior. Morgan's Canon (see Chapter 11: 1894 Morgan's Canon) will almost always lead an investigator to a simpler, less elegant, explanation for an animal's behavior than the use of mental time travel. Simple associations, rather than cognitive planning, for example can

Conceptual Breakthroughs in Ethology and Animal Behavior.
DOI: http://dx.doi.org/10.1016/B978-0-12-809265-1.00037-X

easily explain why a foraging animal would show up at the right time and place to intercept potential prey.

Care needs to be taken not to under-credit animals for their cognitive processes. Just because we can imagine a simpler explanation does not mean that the animal is not engaging in higher-level thinking. Fallacious thinking along these lines sometimes originates with Morgan's canon. Studies in primates and corvids (jays, crows, and their relatives) suggest that these species have abilities for mental time travel (Clayton et al., 2003; Suddendorf and Corballis, 2010). As a very interesting cognitive process, mental time travel remains in intriguing piece of the puzzle of understanding animal cognition.

IMPACT: 6

Tulving's work is widely cited and recognized as pioneering in describing higher-level cognitive processes. While animal tests remain difficult and no neurobiological basis for episodic memory has been found, this concept has impelled a substantial thread in the scientific conversation about cognition.

SEE ALSO

Chapter 52, 1978 Theory of Mind; Chapter 67, 1992 Working Memory; Chapter 72, 1998 Gaze Following; Chapter 74, 2000 Emotion and the Brain.

REFERENCES AND SUGGESTED READING

Clayton, N.S., Bussey, T.J., Dickinson, A., 2003. Can animals recall the past and plan for the future? Nat. Rev. Neurosci. 4, 685–691.

Murray, B., 2003. What makes mental time travel possible? Psychologist Endel Tulving offered a theory on our uniquely human ability to act today based on our past and future. Monitor Psychol. 34, 62; <http://www.apa.org/monitor/oct03/mental.aspx>.

Stuss, D.T., Knight, R.T. (Eds.), 2012. Principles of Frontal Lobe Function, 2nd ed. Oxford University Press, Oxford; New York. <http://dx.doi.org/10.1093/acprof:oso/9780195134971.001.0001>.

Suddendorf, T., Corballis, M.C., 2010. Behavioural evidence for mental time travel in nonhuman animals. Behav. Brain Res. 215, 292–298.

Tulving, E., Thomson, D.M., 1973. Encoding specificity and retrieval processes in episodic memory. Psychol. Rev. 80, 352–373.

Chapter 38

1973 Game Theory

THE CONCEPT

Is life just a game? This question resonates on many levels. For animals, what if they play life like a game in which they have specific strategic and tactical choices? How could their choices be partly genetically programmed and partly contingent on circumstances? Game theory gives us tools to analyze just these sorts of problems.

THE EXPLANATION

I chose Maynard Smith and Price (1973) paper as a watershed moment in the application of game theory to animal behavior. Game theory comes back in Chapter 56, 1981 Prisoner's Dilemma, and remains a valuable approach to considering evolutionary processes that shape animal behavior. The Maynard Smith and Price (1973) paper is also highlighted in Chapter 41, Animal Conflict; it speaks volumes about the impact of this short paper that I chose to use it as the basis for two chapters. There is some overlap between this chapter and Chapter 41, but the themes of the two–game theory in this chapter and conflict in Chapter 41–are distinct. Maynard Smith and Parker (1976) give a more extensive analysis of the use of game theory in evolutionary biology. In a book published a few years later Maynard Smith (1982) gives a very in-depth treatment of the topic.

Games can be fully informed—another way of saying this is that the players have complete information. Chess is an example of a game in which each player has complete information, because both players can see the position of all the pieces on the board. Poker and bridge, in contrast, are partially informed (incomplete information), as each player's cards are concealed from the other players, and both of these games require players to assess their opponents' holdings based on information they can collect in the process of the game. In many animal games, such as mating or competition for resources, concealment or even misrepresentation of information plays an important role, so animals often rely on assessments and probabilities to make correct choices, much in the way that bridge or poker players rely on incomplete information.

Conceptual Breakthroughs in Ethology and Animal Behavior.
DOI: http://dx.doi.org/10.1016/B978-0-12-809265-1.00038-1

In any reasonably informed game (the aforementioned examples of bridge, chess, and poker come to mind among human games, but more physical games like basketball and football also fit this mode) players choose strategies and tactics.[1] A strategy is an overall plan for winning the competition; e.g., a bird may choose a strategy of overtaking a territory held by another bird. Tactics are specific moves designed to help to achieve the overall goal. So our bird might wait until the other bird leaves its territory to feed, and make its move then, or it might engage in an all-out attack.

Maynard Smith and Price (1973) used the Hawk–Dove game for their analysis. In this game an animal may be either a hawk—attacking all the time and risking injury—or a dove—never attacking but also not risking injury. If one imagines grain strewn on a sidewalk and pigeons collecting the grain, then some might be hawks, wresting grain from others, while others might be doves, yielding if challenged. Interestingly enough, pigeons tend to have passive, dove-like strategies. The same set-up but with crabs strewn on a beach reveals that gulls tend to have very aggressive, more hawk-like, strategies. Simple games with easily visualized strategies have extraordinary explanatory power in animal behavior.

Ultimately, given enough generations of selection, evolution centers on what game theorists call an Evolutionarily Stable Strategy, or ESS. An animal that plays perfectly may be unbeatable by currently available strategies, but changing environments or innovations—newly evolved approaches—can destabilize the game and send evolution on course to find a new ESSs. Thus behavioral games are constantly evolving, even if an ESS exists under the current conditions (Fig. 38.1).

IMPACT: 5

Game theory remains an important thread in the development of models for conflict and cooperation in animals. The adaptability of game theory to computer modeling allows investigators to explore the stability of strategies over simulated evolutionary time and to compare the results of virtual behavior with empirical observations. The outcomes are interesting and valuable for testing evolutionary hypotheses.

SEE ALSO

Chapter 40, The Red Queen; Chapter 41, 1973 Animal Conflict; Chapter 56, 1981 Prisoner's Dilemma.

---

1. Examples of games with no information for players and thus having participants who are completely uninformed are slot machines, dice, and roulette. Payoffs are based on chance alone. In casino versions of these games, the odds are tilted to the casino so players always ultimately lose. There is no point in choosing tactics or strategies for these games, other than not playing them at all.

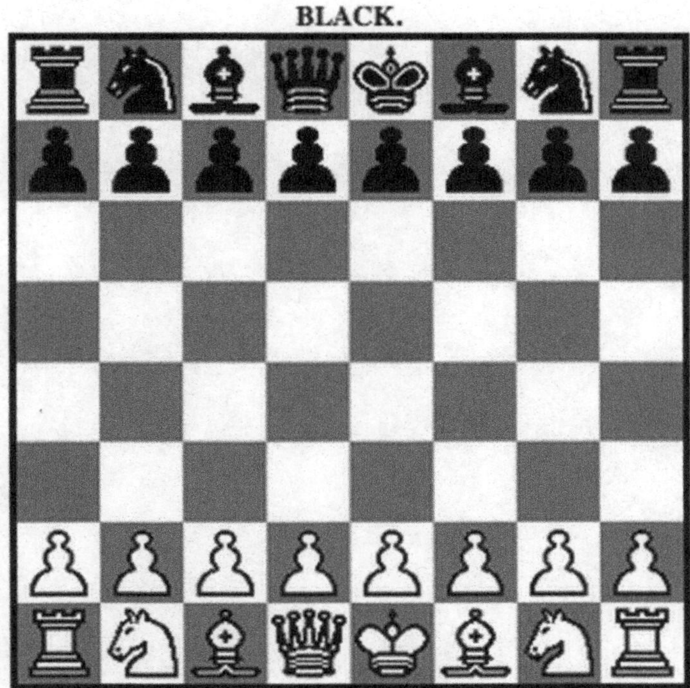

FIGURE 38.1 The starting position in a chess game. Once possession of the black and white pieces is determined, each player can assess the positions of all the pieces, and plot their strategy and tactics accordingly. Each player has the same information, although they may differ in skill and experience. In many evolutionary games animals work with incomplete information; this introduces an element of chance to the outcomes.

REFERENCES AND SUGGESTED READING

Fawcett, T.W., Hamblin, S., Giraldeau, L.A., 2013. Exposing the behavioral gambit: the evolution of learning and decision rules. Behav. Ecol. 24, 2–11.

Johnstone, R.A., Hinde, C.A., 2006. Negotiation over offspring care-how should parents respond to each other's efforts? Behav. Ecol. 17, 818–827.

Maynard Smith, J., 1982. Evolution and the Theory of Games. Cambridge University Press, Cambridge, UK, 234 pp.

Maynard Smith, J., Price, G.A., 1973. The logic of animal conflict. Nature 246, 15–18.

Maynard Smith, J., Parker, G.A., 1976. The logic of asymmetric contests. Anim. Behav. 24, 159–175.

Prestwich, K.N., 1999. A simple game: Hawks versus Doves. <http://college.holycross.edu/faculty/kprestwi/behavior/ESS/HvD_intro.html>.

Chapter 39

1973 The Many Eyes Hypothesis

THE CONCEPT

Being in a group allows sharing of information about predator location and threat level. By having many eyes on sentry duty, the chances of spotting an approaching threat are increased.

THE EXPLANATION

Why do animals come together in groups? Herds, schools and flocks are everyday occurrences in animal behavior. Yet their existence creates puzzles in evolutionary biology. On the surface, a group of animals is a much more concentrated and higher-value food target for a predator than individual animals would be if widely scattered in the habitat. We have already seen, in Chapter 36: 1971 Selfish Herds, that animals in groups can gain value from their association with the group. One of the explanations, the *many eyes hypothesis*, states that each animal needs to invest less in vigilance as group size goes up (Pulliam, 1973). As a corollary, animals should monitor how vigilant their neighbors in the group are, and adjust their behavior correspondingly.

There are downsides to group membership. When in groups animals are open to observation by others of their species, so information about food discoveries, nesting sites and the like becomes public information (see Chapter 79: Public and Private Information). Many parasites and diseases are much more easily passed from animal to animal when they are close together. Living in groups seems a senseless approach if it facilitates disease, so there must be counterbalancing evolutionary forces that keep groups together.

In fact, so many species of animal group together that humans have come up with an entire dictionary of species-specific names for them: a murder of crows (my favorite), a caravan of camels, a parliament of owls, and so on. There must be strong evolutionary pressures that favor living in groups. Evolutionary explanations for the ultimate causes of grouping abound (see Chapter 28: 1964 Inclusive Fitness and the Evolution of Altruism, and Chapter 35: 1971 Reciprocal Altruism, for example). These explanations

Conceptual Breakthroughs in Ethology and Animal Behavior.
DOI: http://dx.doi.org/10.1016/B978-0-12-809265-1.00039-3

focus attention on fitness benefits to the individual members of the group and give us ways to account for how the benefits of group membership might outweigh the costs.

Another approach is to focus on the mechanisms that facilitate information flow within groups. The many eyes hypothesis allows us to consider optimal group size. For a small group, adding another pair of eyes can substantially improve the group's ability to spot approaching predators. Adding eyes also can reduce the vigilance burden for each group member. By incorporating the value of each member of the group in adding vigilance effort, Pulliam's (1973) work allows modeling to predict optimal group size. This approach integrates well with Hamilton's selfish herd concept (see Chapter 36: Selfish Herds) which shows how position within the group affects the costs and benefits of grouping.

Pulliam's (1973) short note came at a critical moment, when investigators may have been leaning too much toward just-so explanations of the advantages of group living. In clear terms, with a simple mathematical model in support, Pulliam showed that the value of shared vigilance for predators, even if each animal contributes equally, rises as the group size increases and then levels off. Larger groups can arise, though, if animals join on the chance that they will achieve more sheltered central positions in the group (see chapter: Selfish Herds).

IMPACT: 2

Following Hamilton's (1971) paper on animal groupings (see chapter: Selfish Herds) and Pulliam's (1973) contribution about flocking, a spurt of literature on the problem of grouping and vigilance presented tests of the reasons why animals come together in groups. Lima's (1995) work is a particularly insightful test, looking at the effect of reduced individual vigilance in some group members, due to hunger, on the responses of other group members.

SEE ALSO

Chapter 28, 1964 Inclusive Fitness and the Evolution of Altruism, and Chapter 35, 1971 Reciprocal Altruism; Chapter 36, Selfish Herds; Chapter 45, 1975 Group Selection.

REFERENCES AND SUGGESTED READING

Hamilton, W.D., 1971. Geometry for the selfish herd. J. Theor. Biol. 31, 295–311.

Lima, S.L., 1995. Back to the basics of antipredatory vigilance-the group-size effect. Anim. Behav. 49, 11–20.

Pulliam, H.R., 1973. On the advantages of flocking. J. Theor. Biol. 38, 419–422.

Treves, A., 2000. Theory and method in studies of vigilance and aggregation. Anim. Behav. 60, 711–722.

Chapter 40

1973 The Red Queen

THE CONCEPT

Evolutionary arms races are self-perpetuating, with each species countering other species' innovations with its own innovations. A predator species that evolves to run faster selects for prey that, in turn, either run faster or are more evasive. Response and counter-response give the appearance of running in place, with neither species making gains over the other.

THE EXPLANATION

In Lewis Carroll's *Through the Looking-Glass* Alice runs with the Red Queen in a race whose beginning and end are never in sight. No matter how fast the Red Queen and Alice run, they seem to be stuck in the same place. This is exactly the quandary presented by evolutionary arms races. No matter how fast a species evolves, the species it interacts with evolve equally quickly. This gives the appearance that evolution is running in place, with no progress made. But stopping is not an option, as then the other species would rapidly win the race.

Carroll's description of Alice's race with the Red Queen caught the attention of evolutionary biologist Leigh Van Valen, who argued in a 1973 paper that "organisms in an adaptive zone … can link the perturbations across time in that the effects of one perturbation may depend on the effects of those before it." In plainer English, evolutionary responses occur in the context of what's already happened and create a context for what will happen. Van Valen's concern was with species extinction, but evolutionary biologists have since applied this principle in predator–prey arms races[1], host–parasite arms races, and the evolution of sexual reproduction.[2]

---

1. "the only requirement for stability of interactions … is that the physical environment not change so as to alter the relative rates at which two species evolve." (Schaffer and Rosenzweig, 1978).
2. Van Valen's mathematical arguments in his original paper were criticized by other leading evolutionary biologists of the time (see Lively (2010) for a review of the controversy). The metaphor of the red queen survives in evolutionary biology, even if the original math has been controversial.

Conceptual Breakthroughs in Ethology and Animal Behavior.
DOI: http://dx.doi.org/10.1016/B978-0-12-809265-1.00040-X

It is fun to consider the fuller context of the Red Queen from *Through the Looking-Glass*, the actual quotation used by Van Valen is in bold:

The most curious part of the thing was, that the trees and the other things round them never changed their places at all: however fast they went, they never seemed to pass anything. "I wonder if all the things move along with us?" thought poor puzzled Alice. And the Queen seemed to guess her thoughts, for she cried, "Faster! Don't try to talk!"....

"Nearly there!" the Queen repeated. "Why, we passed it ten minutes ago! Faster!" And they ran on for a time in silence, with the wind whistling in Alice's ears, and almost blowing her hair off her head, she fancied. ...

The Queen propped her up against a tree, and said kindly, "You may rest a little now."

Alice looked round her in great surprise. "Why, I do believe we've been under this tree the whole time! Everything's just as it was!"

"Of course it is," said the Queen, "what would you have it?"

"Well, in our *country," said Alice, still panting a little, "you'd generally get to somewhere else—if you ran very fast for a long time, as we've been doing."...*

*"A slow sort of country!" said the Queen. "**Now,*** **here***, **you see, it takes all the running*** **you** ***can do, to keep in the same place**. If you want to get somewhere else, you must run at least twice as fast as that!"*

The famous phrase "the hurrier I go the behinder I get," which describes the Red Queen's race, doesn't actually appear in Lewis Carroll's book, but is very much a part of our general understanding of the race.

In practical terms, evolution is limited by the necessity for genetic variation in the trait under selection. Evolution is the change in gene frequencies in a population over time, and without genetic variation there can be no evolution. Physical and mechanical constraints imposed by nature also constrain evolution. Finally, trade-offs shape evolution; if a species evolves to deal with one type of threat or environmental need, that evolution may have a cost in making the species unable to meet other threats or needs. Over the vast expanse of evolutionary time, mutation and possibly introgression of genes from other species can add genetic variation with which evolution can work, but at any given point in evolutionary time if a new threat emerges and a species lacks the genetic variation to evolve its response, then it falls behind in the arms race. Species introduced to North America, such as the starling, the English sparrow, and the imported fire ant find relatively little resistance to their spread because they appear to be ahead in evolutionary arms races in which native species have no immediately available responses.

IMPACT: 7

This compact and attractive analogy has helped generations of students to understand the principles of coevolution.

SEE ALSO

Chapter 38, 1973 Game Theory.

REFERENCES AND SUGGESTED READING

Abrams, P.A., 2000. The evolution of predator-prey interactions: theory and evidence. Ann. Rev. Ecol. Syst. 31, 79–105.

Dodgson C. AKA Lewis Carroll. The Project Gutenberg EBook: Through the Looking-Glass [EBook #12].

Lively, C.M., 2010. A review of red queen models for the persistence of obligate sexual reproduction. J. Hered. 101, S13–S20.

Lively, C.M., Dybdahl, M.F., 2000. Parasite adaptation to locally common host genotypes. Nature 405, 679–681.

Schaffer, W.M., Rosenzweig, M.L., 1978. Homage to red queen 1. Coevolution of predators and their victims. Theor. Population Biol. 14, 135–157.

Schmid-Hempel, P., Ebert, D., 2003. On the evolutionary ecology of specific immune defence. Trends Ecol. Evol. 18, 27–32.

Van Valen, L., 1973. A new evolutionary law. Evol. Theo. 1, 1–30.

Chapter 41

1973 Animal Conflict

THE CONCEPT

Conflict among animals, whether characterized as dominance hierarchy, aggression, or as agonistic behavior, can be explained by game theory. Evolution drives the behavior of animals in conflict in ways that maximize their reproductive potential, even though sometimes that leads to risk of injury or death. Ritualized encounters, in which risk of injury is reduced, also play an important role in animal conflict. Game theory helps us to understand the nature of animal conflict and how outcomes are shaped.

THE EXPLANATION

I give this concept a separate chapter because of the importance of understanding why animals often do not fight to the death, but why sometimes they do. Game theory gives us tools for analyzing the strategies and tactics used by animals in contests for resources (Maynard Smith and Price, 1973). This chapter stems from the same paper as Chapter 38, Game Theory; here the focus is more on the evolution of animal conflict.

For example, game theory can consider conflicts in which animals have strategies of escalation—proceeding to more violent acts—or de-escalation available to them. Escalation may yield a win but is also is more risky. De-escalation hardly ever wins but is much less risky. Sometimes de-escalation in a given encounter is the best strategy as it allows the animal to live to fight another day.

The escalation/de-escalation choice leads to one of the simplest games, *Hawk versus Dove*. Hawks always escalate and Doves never escalate. If Hawks only encounter Doves, then they always win, but in a mixed population they will sometimes encounter other Hawks and may die, losing all potential reproduction. Doves coexist but never have the opportunity to monopolize resources; maximum payoffs are always less for Doves than for Hawks. By viewing conflict with game theory, the investigator can deduce the *Evolutionarily Stable Strategies (ESS)* for the animals; these strategies continue to exist over evolutionary time because they will never be fully

Conceptual Breakthroughs in Ethology and Animal Behavior.
DOI: http://dx.doi.org/10.1016/B978-0-12-809265-1.00041-1

replaced by another strategy. Under some circumstances, Hawk and Dove are evolutionarily stable strategies.

Part of the background to Maynard Smith and Price's paper is that at the time of its publication among nonscientists there was a generally held belief that animals would behave for "the good of the species." This included the noble but incorrect view that non-human animals never fight to the death. Reports of lethal fighting among animals were dismissed as aberrations. The existence of Hawks as an evolutionarily stable strategy in the Hawk versus Dove game shows that fighting to the death is a viable evolutionary option, and a less biased view of natural history shows that, in fact, lethal fights occur in many species. A Hawk versus a Hawk may very well be a fight to the death.

An evolutionary alternative to out and out fighting is ritualization. Again, game theory (Maynard Smith and Price, 1973; Maynard Smith and Parker, 1976) helps us to understand that displays can symbolically represent an animal's potential strength in a conflict. Signals of intent and strength include a cock crowing, an elk bugling, or a peacock flaring its train. These are ritualized symbols of what the animal might do. Ritualization allows combatants to assess relative strengths without entering the complete combat of the Hawk strategy in Hawk versus Dove.

An interesting sidelight to this is that in most territorial contests the possessor of the territory has an advantage—is most likely to win—even if objectively it is the weaker animal. This behavioral home field advantage can be quite significant in predicting outcomes of contests. For an intruder, yielding may be better than the risks of fighting in unfamiliar terrain. Alternatively, possession of territory may give the holder stronger motivation to fight.

To answer the question posed at the beginning of this chapter, animals do fight to the death. They do so when the fitness stakes are very high, particularly when the looser might not have much fitness potential even if they survive. In a population with a mix of Hawks and Doves, Hawk–Hawk encounters are not likely to end well for one of the combatants (Fig. 41.1).

FIGURE 41.1 Bighorn sheep clashing. This is a game of strength, endurance, and injury avoidance. The male bighorn's skull has evolved to absorb the shock and prevent brain injury, but these contests are at the limits of what each animal can survive.

IMPACT: 3

These papers, first Maynard Smith and Price (1973) and then Maynard Smith and Parker (1976), brought game theory to the forefront as a tool for analyzing animal contests. The publications helped animal behaviorists develop a rationale for understanding risk of injury and death in animal contests and for countering the public perception that animals would behave selflessly so that the species might survive.

SEE ALSO

Chapter 38, 1973 Game Theory; Chapter 56, 1981 Prisoner's Dilemma.

REFERENCES AND SUGGESTED READING

Maynard Smith, J., Price, G.R., 1973. The logic of animal conflict. Nature 24, 15–18.

Maynard Smith, J., Parker, G.A., 1976. The logic of asymmetric contests Anim. Behav 24, 159–175.

Prestwich K.N., 1999. A simple game: Hawks versus Doves. <http://college.holycross.edu/faculty/kprestwi/behavior/ESS/HvD_intro.html>.

Riechert, S.E., 2013. Maynard Smith & Parker's (1976) rule book for animal contests, mostly. Anim. Behav. 86, 3–9.

Chapter 42

1974 *Caenorhabditis elegans* Behavioral Genetics

THE CONCEPT

All of the genes in a species can be known and their roles determined, and this can specifically be applied to analysis of behavioral phenotypes.

THE EXPLANATION

Caenorhabditis elegans is a tiny roundworm. It has a simple internal anatomy, limited sensory abilities, and a very constrained nervous system. Yet we know more about this worm's biology and how its development, physiology and behavior relate to its genetics than any other animal (Brenner, 1974). Remarkably more. The fruit fly, *Drosophila melanogaster*, is a distant second in terms of how well we comprehend these interrelationships. Current knowledge about *C. elegans* is documented in an open access online resource, *The Worm Book* (http://www.wormbook.org/).

Caenorhabditis elegans has six pairs of chromosomes, including a pair of sex chromosomes and about 100 million base pairs in its genome. It was the first eukaryotic genome to be sequenced and it still stands unique as the most thoroughly sequenced and annotated genome. *Caenorhabditis elegans* has two sexes, males, which have 959 somatic (body) cells, and hermaphrodites, which have 1031 somatic cells. The developmental origin, role, and ultimate fate of each cell has been mapped, so that we know that the nervous system contains exactly 302 neurons and 56 glial (support) cells. In other words, roughly one-third of the cells in the body are devoted to the nervous system.

Such detailed developmental and anatomical knowledge of the nervous system means that the neurobiology of behavior can be studied in minute detail. Even though these are small and relatively simple animals, they forage, search for good habitat, choose mates, and enter a resting state called the dauer, which allows them to survive unfavorable conditions. There is little in the normal roster of animal behavior, except problem solving and

Conceptual Breakthroughs in Ethology and Animal Behavior.
DOI: http://dx.doi.org/10.1016/B978-0-12-809265-1.00042-3

complex memory-based navigation, that *C. elegans* does not accomplish on some level.

For example, there are social and solitary strains of *C. elegans*, and we know quite a bit about the behavioral and genetic differences between the strains. Intriguingly, Bono and Bargmann (1998) found that a change at a single locus, which codes for a neuropeptide receptor, could switch animals between the social and solitary foraging modes. This type of result allows deeply reductionist exploration of how behavioral phenotypes are shaped by gene expression.

While genomic sequencing has made genomes available for many animal species, the depth of knowledge about *C. elegans* and the simplicity of its nervous system still make it an outstanding model for studies that combine behavior, genetics, and neuroscience.

IMPACT: 9

This paper on *C. elegans* (Brenner, 1974) comes relatively early in the molecular revolution in genetics and it establishes the groundwork upon which much of the subsequent knowledge is based. Brenner found mutations in the *C. elegans* genome, mapped those mutations, and explored their functional effects. His work demonstrated that *C. elegans* could serve as an excellent model system, standing at just the right point of having enough complexity without being overwhelmingly complicated.

SEE ALSO

Chapter 34, 1971 Behavioral Genetics; Chapter 75, 2000 Social Amoebas and Their Genomes.

REFERENCES AND SUGGESTED READING

Benzer, S., 1971. From the gene to behavior. J. Am. Med. Assoc. 218, 1015–1022.

Bono, M., Bargmann, C.I., 1998. Natural variation in a neuropeptide Y receptor homolog modifies social behavior and food response in *C. elegans*. Cell 94, 679–689.

Brenner, S., 1974. Genetics of *Caenorhabditis elegans*. Genetics 77, 71–94.

Chapter 43

1974 Standardizing Behavioral Observation Methods

THE CONCEPT

In the early years of animal behavior studies, the most commonly used methodology was ad hoc observation, with the investigator picking and choosing the animals that seemed most interesting and then recording behavior that drew the investigator's attention. This was adequate to gain basic natural history knowledge about species, but did not allow for testing hypotheses or for making quantitative comparisons. Rather late in the development of the field, Tinbergen framed his four questions (see Chapter 26: 1963 The Four Questions). Only a decade after Tinbergen's publication, Altmann (1974) provided a roadmap for making systematic quantitative behavioral observations.

THE EXPLANATION

Jeanne Altmann became famous for her field studies of baboons at the Amboseli National Park in Kenya. The Amboseli project, much like Jane Goodall's work on chimpanzees in the Gombe National Park in Nigeria, relies on following the behavior of animals through their lifetimes, as well as across generations. By looking at how the behavior of one generation casts a reflection on the behavior of subsequent generations scientists gain a deep understanding of culture and social behavior.

But if a study spans decades, many different scientists, students, and technicians will have been involved in the collection of data. If the data from the first decade of the study can't be compared to data from the third decade, then much of the value of the research program is lost.

Jeanne Altmann was concerned enough about standardization of observation techniques that in 1974 she published a paper that captures best practices for watching animals and recording their behavior (Altmann, 1974). This paper has been cited over 11,000 times and is still the go-to resource for advice about designing studies of animal behavior. Altmann's genius was

Conceptual Breakthroughs in Ethology and Animal Behavior.
DOI: http://dx.doi.org/10.1016/B978-0-12-809265-1.00043-5

in realizing that the key to unbiased sampling was having a predetermined plan for how to distribute observations in a population of animals.

Here's a summary of how most field biologists do their studies of animal behavior:

In most behavioral studies conducted under field conditions the first step is construction of an *ethogram*, a list of the observed behaviors with definitions and perhaps drawings or photos. These seemingly mundane starting points opens the door for accurately measuring how often animals perform each behavior (Anon, no date). An ethogram can serve as the basis for a *time budget* for the animals: How much time is each animal spending on each activity? Time budgets have and important and substantial role in current research in animal behavior as they allow us to test hypotheses by making comparisons. Do, for example, birds spend more time calling when urban traffic noise is high?

In most studies, the first step after defining hypotheses is to plan observations so that data is collected in an unbiased and accurate manner. Alternatives include:

Focal animal sampling: Choosing a specific animal to focus on, ensuring accurate records of that animal's behavior. The focal animal can be changed after a period of time so that the sample size is increased.

Ad libitum sampling: The behavior of animals is randomly sampled, optimally using a random numbers generator to determine which animal is being observed at a given point in time.

Sequential sampling: Sampling attention moves from animal to animal as they execute a sequence of behaviors.

Instantaneous/scan sampling: The behavior of all animals in view is recorded at a given instant.

The sampling technique is chosen to best accomplish the major goal of collecting data that will test the scientist's hypotheses. For example, if the hypothesis is that other animals groom dominant females more often than subordinate females, then we can first consult the ethogram to find dominance and grooming behaviors. Next we observe behaviors in our group of animals to find which females are dominant over other group members. Finally, we would choose dominant and subordinate females as *focal animals* and we would watch each of our focal animals for equal periods of time, and record grooming behavior.

This seems simple enough, but there's a lot going on with this approach to designing behavioral studies. Foremost, in any animal behavior research, observations should not be skewed by the behavior of one animal. We know that individuals can be idiosyncratic, and if observations are biased by an atypical animal then the whole thing is likely a waste of time. Thus the unbiased choice of animals to observe and the observation of enough animals to account for variation among animals in their behavior is of the utmost importance. The program outlined by Altmann (1974) is the pathway to excellence in animal behavior research.

IMPACT: 9

Altmann's highly cited work promoted the systematic collection of field data on animal behavior. Students still use her paper as a guidebook for experimental design and unbiased sampling.

SEE ALSO

Chapter 26, The Four Questions.

REFERENCES AND SUGGESTED READING

Altmann, J., 1974. Observational study of behavior: sampling methods. Behaviour 49, 227–267.

Anon, no date. This link leads to an online chimpanzee ethogram, with written descriptions and drawings: <http://gombechimpanzees.org/activities/behavior-research/>.

Bell, A.M., Hankison, S.J., Laskowski, K.L., 2009. The repeatability of behaviour: a meta-analysis. Anim. Behav. 77, 771–783.

Dawkins, M.S., 2007. Observing Animal Behaviour: Design and Analysis of Quantitive Controls. Oxford University Press, 168 pp.

Martin, P., 2007. Measuring Behavior: An Introductory Guide, third ed. Cambridge University Press, 186 pp.

Chapter 44

1974 Parent–Offspring Conflict

THE CONCEPT

Evolution typically pushes parents to invest just enough in each of their offspring to ensure survival and reproduction, but parents also often conserve resources for future offspring. The offspring, on the other hand, are more driven by evolution to demand investment that will enhance their own chances, even at the expense of their parents and their siblings. This situation leads to parent–offspring conflict over how much parents are willing to invest in their offspring.

THE EXPLANATION

That parents and their offspring could have conflict is no surprise to any human; parent–offspring conflict is part of the texture of many families' lives and certainly offers juicy material for soap operas and movies. Trivers (1974) made two important contributions to comprehending why parent animal conflict is widespread in animals. First, he established a general framework for knowing how long and how much parents should continue to invest in their offspring. Second, he helped to build new ways of seeing the balance between investment in current offspring and potential future offspring, including how the sex of those future offspring affects current conflicts.

Not surprisingly, Trivers' model predicts parent–offspring conflict increases with the age of the offspring, as the balance shifts from what is absolute necessity for offspring survival to continued maintenance of offspring that may be able to care for themselves. During this later phase of development, weaning conflict occurs and in Trivers' characterization, offspring may use "psychological" weapons in gaining more parental support (see Chapter 66: 1991 Receiver Psychology). While this seems to be an anthropocentric way of interpreting animal behavior, for parents there is clearly a clash between the physiological mechanisms that have driven bonding with offspring and the parent's own needs. Mammals and birds have similar-enough hormonal drivers of parental care to support the assertion that

Conceptual Breakthroughs in Ethology and Animal Behavior.
DOI: http://dx.doi.org/10.1016/B978-0-12-809265-1.00044-7

FIGURE 44.1 Even in seemingly placid situations like this Canada goose family, differences in evolutionary interests can create conflict between parents and offspring. *Photo courtesy of Michael Breed.*

the emotional bond between parents and offspring can cause the parent to overinvest (see Chapter 64: 1990 The Challenge Hypothesis).

Trivers' (1974) analysis also showed that the degree of genetic relationship between offspring could shift patterns of parental investment. While it seems, from a mammalian perspective, obvious that all offspring are equally related, in some animals, such as wasps, bees, and ants (the insect order Hymenoptera), males arise from unfertilized eggs, ultimately yielding a situation in which full sisters are more genetically similar than in animals like mammals or birds. This changes the value of future sisters to current female offspring and makes conflict over parental investment less likely.

Maynard Smith (1977) followed on Trivers' publication with a game theory analysis of how mating systems options for animals interact with parent–offspring conflict. Maynard Smith (1977) chides Trivers for not having expressed the cost of parental care in terms of lost future reproduction. However, the overall result of Maynard Smith's models is supportive of the conclusions drawn by Trivers (Fig. 44.1).

IMPACT: 9

The enduring conversation about parent–offspring conflict and the strategies that both sides employ to maximize their evolutionary outcomes has been fascinating. Innovations in genetic techniques have allowed incorporation of certainty of parenthood in to the models (see Chapter 50: 1977 The Evolution of Mating Systems, for a discussion of extrapair copulations). Technical advances in physiology and biochemistry have allowed more detailed assessments of nutrition gained from parents. Finally, social network analysis allows assessment of the reverberations of parent–offspring interactions through the social bonds formed by the offspring through their lifetime. All of these new approaches are beginning to come together to allow comprehensive analyses of parent–offspring conflicts and their consequences.

SEE ALSO

Chapter 29,1965 Harry Harlow and Social Isolation in Monkeys; Chapter 38, 1973 Game Theory; Chapter 50, 1977 The Evolution of Mating Systems; Chapter 53, 1980 Dispersal.

REFERENCES AND SUGGESTED READING

Emlen, S.T., 1995. An evolutionary theory of the family. Proc. Natl. Acad. Sci. USA 92, 8092–8099.

Godfray, H.C.J., 1991. Signaling of need by offspring to their parents. Nature 352, 328–330.

Godfray, H.C.J., 1995. Evolutionary-theory of parent-offspring conflict. Nature 376, 133–138.

Houston, A.I., Székely, T., McNamara, J.M., 2013. The parental investment models of Maynard Smith: a retrospective and prospective view. Anim. Behav. 86, 667–674.

Maynard Smith, J., 1977. Parental investment: a prospective analysis. Anim. Behav. 25, 1–9.

Trivers, R.L., 1974. Parent-offspring conflict. Am. Zool. 14, 249–264.

Chapter 45

1975 Group Selection

THE CONCEPT

Animals might behave in ways that are detrimental to themselves if their behavior enhances the survival of the group in which they live. This model does not require that the animals in the group be close relatives. Group selection is an alternative to kin selection in explaining the evolution of altruism.

THE EXPLANATION

One of the great fights among scientists who study the evolution animal behavior centers on how to explain the existence of altruism. We've already visited this topic in Chapter 28, 1964 Inclusive Fitness and the Evolution of Altruism and Chapter 35, 1971 Reciprocal Altruism.[1]

David Sloan Wilson, in 1975, took a very different view of how altruism could evolve in animal populations, arguing that group selection could be a viable force for evolution Wilson (1975). He felt that genes for altruism could spread in populations if altruistic populations survived longer than selfish populations. Imagine a number of small subpopulations, or demes, of an animal species. In this circumstance most of the mating in the species goes on within the demes. There is competition among animals in each deme, but also the potential for cooperation. If genes for cooperation arise within demes and then the demes with those genes have higher survival rates than demes without the genes, it is conceivable that selection at the group (deme) level could favor the persistence of the gene for cooperation.

Wynne Edwards (1962, 1986) had another interesting view of altruism and cooperation in animals. He argued that animals could behave for the good of their species. For example, in his model an animal might reduce its

1. Altruistic behavior has a cost for the donor and a benefit for the recipient. In Darwinian terms, we do not expect animals to pay a fitness cost unless they receive as much or more fitness benefit in return. Inclusive fitness (kin selection) postulates that the benefit comes from increased reproduction of close relatives, so that the donor's genes are passed to the next generation. Reciprocal altruism invokes an arrangement in which the donor later becomes the beneficiary—a kind of contract for reciprocated helping between the two animals.

Conceptual Breakthroughs in Ethology and Animal Behavior.
DOI: http://dx.doi.org/10.1016/B978-0-12-809265-1.00045-9

own reproduction in order to save others from starving, or animals might not fight to the death as that would be a negative outcome for the species. This is a bit contrary to the Darwinian logic of natural selection and animals probably never behave "for the good of the species." Wynne Edwards' work comes up here because the good of the species argument is often loosely linked with group selection.

More recently Nowak et al. (2010) revived the group selection idea, presenting mathematical models that purported to show that group selection, much as described by Wilson (1975) explain the evolution of sterile worker castes in social insects, as well as other altruistic social phenomena. This led to a storm of dissent, notably a critique signed by over 100 scientists (Abbott et al., 2011). The inclusion of E. O. Wilson as an author on the Nowak et al. (2010) paper was consistent with his long-term negative view of kin selection (expressed in his 1975 book, *Sociobiology*). E. O. Wilson is clearly an outlier in his views on this topic. Queller et al. (2015) have provided a data-rich critique of the Nowak et al. (2010) paper and, indeed, the data overwhelmingly supports kin selection as the driving mechanism in the evolution of social behavior (Fig. 45.1).

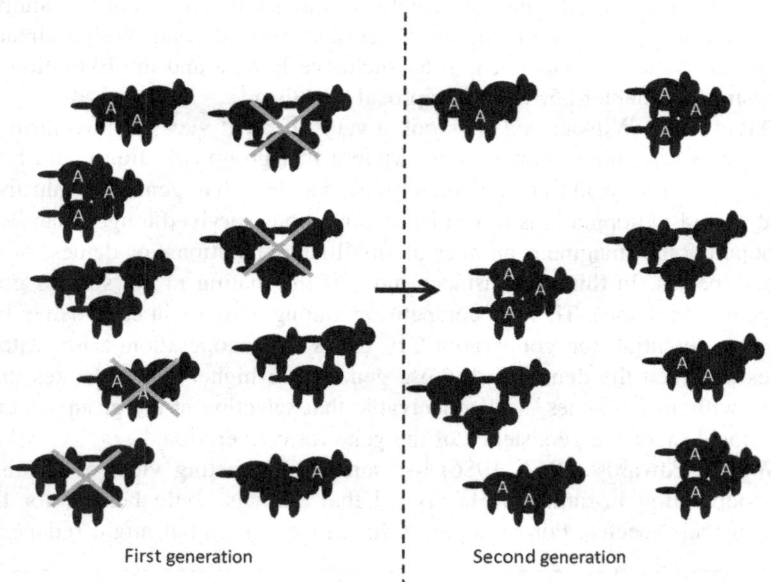

FIGURE 45.1 In the figure on the left, some social groups (demes) contain animals with an altruistic gene (A). Those groups survive and reproduce more often than groups without the altruistic gene. Failure to survive and reproduce is indicated by the X through the group. On the right, the second generation of this population is illustrated. Note that more groups contain altruists (A), reflecting the better reproductive rate of the A groups in the first generation. This model does not rely on kinship to drive selection for the altruistic gene.

IMPACT: 2

There's more smoke than fire here, as Wilson's (1975) model requires a very specific set of population circumstances that remains undocumented in nature. The Nowak et al. (2010) models are riddled with unrealistic assumptions and while they certainly generated controversy, the vast majority of empirical evidence on altruism supports kin selection as the evolution explanation for such behavior (Abbott et al., 2011; Strassmann and Queller, 2014; Queller et al., 2015).

SEE ALSO

Chapter 28, 1964 Inclusive Fitness and the Evolution of Altruism; Chapter 35, 1971 Reciprocal Altruism; Chapter 36, Selfish Herds; Chapter 41, 1973 Animal Conflict; Chapter 56, 1981 Prisoner's Dilemma.

REFERENCES AND SUGGESTED READING

Abbott, P., et al., 2011. Inclusive fitness theory and eusociality. Nature 471 (7339), E1–4.

Liao, X., Rong, S., Queller, D.C., 2015. Relatedness, conflict, and the evolution of eusociality. PLOS Biol. 13, e1002098.

Nowak, M.A., Tarnita, C.E., Wilson, E.O., 2010. The evolution of eusociality. Nature 466, 1057–1062.

Queller, D.C., Rong, S., Liao, X., 2015. Some agreement on kin selection and eusociality? PLOS Biol. 13, e1002133.

Strassmann, J.E., Queller, D.C., 2014. Kin selection Hamiltonian biology, from microbes to god. Anim. Behav. 92, 239–240.

Wilson, D.S., 1975. A theory of group selection. Proc. Natl. Acad. Sci. USA 72, 143–146.

Wilson, E.O., 1975. Sociobiology: The New Synthesis. Belknap Press of Harvard University Press, Cambridge, MA, 720 pp.

Wynne-Edwards, V.C., 1962. Animal Dispersion in Relation to Social Behaviour. Oliver & Boyd, Edinburgh, 653 pp.

Wynne-Edwards, V.C., 1986. Evolution Through Group Selection. Blackwell Scientific, Oxford, 398 pp.

Chapter 46

1975 Sociobiology

THE CONCEPT

All behavior, animal and human, can be interpreted using a unified framework based on natural selection. Genes and environment both play a role in determining behavior, but genetic hypotheses for social phenomena should always be considered.

THE EXPLANATION

Wilson's (1975) book *Sociobiology: The New Synthesis* created far more than a splash in the pond of academic biology. In fact the repercussions were more like a tidal wave. In a magnificent sweep, he integrated much of what was currently known about animal behavior with ecological and evolutionary theory. Never one to demur, Wilson claimed publicly to have invented a new field of inquiry and that the synthesis would stand as a turning point in how we understand animals and in how humans understood each other.

Few of the biological facts in Wilson's book created controversy. The gathering of so much information about animal behavior and the coverage of the entire diversity of animals made this book an important reference for students of behavior, and to this day it retains that value.

However, Wilson's insistence in this book, in subsequent books, and in public statements, for the strength of genetic and evolutionary influences on human behavior and human culture revived the nature/nurture debate that had raged off and on since the end of World War II. In the narrowest of scientific terms this argument started with a clash between the European ethologists—Lorenz, Tinbergen, and their followers—and the American comparative psychologists led by Daniel Lehrman (see Chapter 20: 1953 The Chasm Between Ethology and Comparative Psychology). The argument really focused on whether instinct (the ethologists' point of view) drove behavior or if learning was the driver (the comparative psychology point of view). While the debate had generated heat, Lehrman and Lorenz were on friendly terms and the conversation lived in the sphere of ideas.

Conceptual Breakthroughs in Ethology and Animal Behavior.
DOI: http://dx.doi.org/10.1016/B978-0-12-809265-1.00046-0

A much darker turn to the debate peaked in the 1980s when some factions came to associate arguments about genetic differences in behavior with racism and cultural determinism. The Harvard evolutionary biologist R. C. Lewontin and his colleagues (e.g., Gould and Lewontin, 1979; Lewontin et al., 1985) wrote scathing critiques of sociobiology, to which E. O. Wilson responded with equally inflamed language.[1] At a meeting of the American Association for the Advancement of Science (AAAS), water was dumped on Wilson's head by an infuriated group of opponents to sociobiology. The sociobiology debate extended into academe far beyond biology; sociologists, philosophers, students of religion, and political scientists all joined the fray.

Generally speaking, biologists have moved to using the term "behavioral ecology" to capture the nonhuman segment of behavioral biology that Wilson covered in his 1975 book. Contemporary biologists rarely refer to themselves as sociobiologists, probably to avoid the negative social connotations that have been loaded onto that word. Numerous studies published over the decades since Wilson's book focus on the evolution of human behavior, and interesting areas such as perception of attractiveness, mate choice, dietary choice, cooperation and spite have been analyzed with an eye to how genes and environment come together to shape human behavior.

The maelstrom started by Wilson's book has died down. From a biological point of view there is no question that behavioral phenotypes are the product of interactions between an animal's genes and its environment, and that the degree of malleability of a behavioral phenotype depends on the particular behavior and animal species under study. Humans are animals and it's no surprise that this general principle applies as much to humans as to any other animal. Probably more was gained from having this debate engage such a broad segment of academia as well as the general public than was lost due to the intemperate discourse that poisoned parts of the conversation.

IMPACT: 3

The publication by Wilson of *Sociobiology* initiated an acrimonious and intensely personal debate within the scientific community, as well as among scholars in fields as diverse as philosophy, sociology, and political science. My view is that by needlessly clinging to the idea that they could solve the ills of society, some biologists damaged the credibility of scientific inquiry into animal behavior. Outside of science, some of the public became fascinated with behavior through reading *Sociobiology* but others came to question the motives of the scientists making sociobiological

1. Wilson and Lewontin were in the same department at Harvard; this could have made for interesting faculty meetings, but I don't know of any accounts that have been published of their encounters on campus.

arguments. The most positive outcome was the stronger inclusion of evolutionary thinking in interpretations of animal behavior.

SEE ALSO

Chapter 9, 1859 Darwin and Social Insects; Chapter 28, 1964 Inclusive Fitness and the Evolution of Altruism; Chapter 35, 1971 Reciprocal Altruism; Chapter 36, Selfish Herds; Chapter 45, 1975 Group Selection.

REFERENCES AND SUGGESTED READING

Caplan, A.L. (Ed.), 1978. The Sociobiology Debate. Readings on Ethical and Scientific Issues. Harper and Row, New York, 514 pp.

Catherine, D., 2013. In: Zalta, E.N. (Ed.), Sociobiology. The Stanford Encyclopedia of Philosophy, <http://plato.stanford.edu/entries/sociobiology/>.

Gould, S.J., Lewontin, R., 1979. The spandrels of San Marco and the panglossian paradigm: a critique of the adaptationist programme. Proc. R. Soc. Lond., B, Biol. Sci. 205, 581–598.

Kitcher, P., 1985. Vaulting Ambition. MIT Press, Cambridge MA, 470 pp.

Lewontin, R.C., Rose, S., Kamin, L.J., 1985. Not in Our Genes: Biology, Ideology and Human Nature. Pantheon, New York.

Sociobiology Study Group of Science for the People, 1976. Sociobiology: another biological determinism. BioScience 26 (182), 184–186.

Wilson, E.O., 1975. Sociobiology: The New Synthesis. Belknap Press of Harvard University Press, Cambridge, MA, 720 pp.

Wilson, E.O., 1978. On Human Nature. Harvard University Press, Cambridge, MA, 288 pp.

Wilson, E.O., 1984. Biophilia. Harvard University Press, Cambridge, MA, 176 pp.

Wilson, E.O., 1998. Consilience: The Unity of Knowledge. Knopf: Distributed by Random House, New York, 382 pp.

Wilson, E.O., 2013. The Social Conquest of Earth. Liveright, 352 pp.

Wilson, E.O., 2015. The Meaning of Human Existence. W. W. Norton, 208 pp.

Chapter 47

1975 The Handicap Principle

THE CONCEPT

Creating sexual ornamentation comes with a cost, or handicap. This cost enforces a level of honesty in sexual signaling.

THE EXPLANATION

Selection favors any characteristic that yields more offspring for the animal carrying the trait. Suppose females of a species prefer males that can win fights with other males. In this case, fighting ability doesn't need to represent any other advantage in life, such as being a good forager or a good parent; the mere fact that good fighters have more offspring makes them preferable mates. Evolution should favor males who represent themselves as good fighters, and we can imagine that the antlers of a deer or an elk could come to have this sort of value as a signal.

But in this description, what's to keep a deer or an elk from misrepresenting its fighting ability? If females choose among males based on antler size, then selection would simply favor larger and larger antlers. In such cases, after a few generations of evolution, the size of the antlers loses its relationship to fighting ability. *Dishonest signaling* is a key piece of the interplay between the sexes in sexual selection. A signal that can be cheaply produced and which doesn't create risk for its bearer could be the ideal mate attractant, acting like a wad of dollar bills with a single hundred-dollar bill visible on the outside.

Amotz Zahavi's (Zahavi, 1975) great insight was that for a signal used in mate choice to remain honest in the face of evolutionary pressure for dishonesty, the receiver of the signal must demand that the signal be costly to produce or carry. He referred to this as the *handicap principle*; the sender must pay a price and the receiver assesses not just the signal but also the cost of the signal. Zahavi presented his ideas with neither genetics nor mathematics to support them, and a great hubbub ensued among evolutionary biologists, some arguing that the genetics of a costly signal model were unworkable. The theoretical biologist Alan Grafen (1990) laid these concerns

Conceptual Breakthroughs in Ethology and Animal Behavior.
DOI: http://dx.doi.org/10.1016/B978-0-12-809265-1.00047-2

FIGURE 47.1 The cost of carrying heavy ornamentation, such as the antlers of this elk, can make the signal a true measure of the strength and endurance of an animal. *Photo courtesy Lauren Benedict.*

to rest by developing game theory models in which costly signaling strategies emerged as being evolutionarily stable.[1]

Costly signals help to explain why signals used in mate choice evolve within limits. At some point in the course, sexual selection cost kicks in and a limit on the trait is established (Fig. 47.1).

What if there is no cost? Evolutionary biologists call characteristics that result from mate choice operating without a signal cost runaway traits. Given enough evolutionary time, the magnitude of a runaway trait is limited only by genetic constraints on development. The famous statistical biologist R. A. Fisher championed the idea of runaway sexual selection in the first three decades of the 20th century. In Fisher's construct, mate choice and sexual selection could essentially capture a trait and lead to ever-more elaborate signals, although cost ultimately should place a limit on runaway processes. Zahavi's (1975) insight was important because it focused attention on the costs of signals at a critical point in the conversation about mate choice.

IMPACT: 3

The handicap principle helps to explain ridiculous ornamentation in animals and other costly signals in animals. It is a concise and easily understood concept that has helped to shape the conversation on mate choice over the last three decades.

1. Game theorists call an evolutionarily stable strategy an ESS. In a given circumstance for multiple strategies to be ESSs, the stability comes from no strategy being able to completely eliminate all the others.

SEE ALSO

Chapter 8, 1859 Darwin and Behavior; Chapter 50, 1977 The Evolution of Mating Systems; Chapter 58, 1982 The Hamilton–Zuk Hypothesis.

REFERENCES AND SUGGESTED READING

Folstad, I., Karter, A.J., 1992. Parasites, bright males, and the immunocompetence handicap. Am. Nat. 139, 603–622.

Grafen, A., 1990. Biological signals as handicaps. J. Theor. Biol. 144, 517–546.

Iwasa, Y., Pomiankowski, A., 1991. The evolution of costly mate preferences 2. The handicap principle. Evolution 45, 1431–1442.

Johnstone, R.A., 1995. Sexual selection, honest advertisement and the handicap principle-reviewing the evidence. Biol. Rev. Camb. Philos. Soc. 70, 1–65.

Zahavi, A., 1975. Mate selection—a selection for handicap. J. Theor. Biol. 53, 205–214.

Zahavi, A., 1977. Cost of honesty (further remarks on handicap principle). J. Theor. Biol. 67, 603–605.

Zahavi, A., 1999. The Handicap Principle: A Missing Piece of Darwin's Puzzle. Oxford University Press, Oxford, 304 pp.

Chapter 48

1976 Marginal Value Theorem

THE CONCEPT

Animals forage in order to maximize their net return from the environment. In patchy habitats, the marginal value theorem helps to predict when animals will depart a food patch and search for new foraging areas.

THE EXPLANATION

Plants and animals are not randomly scattered in an environment. Generally speaking they occur in patches that align with preferred habitats. Competition for space with a patch may lead to territoriality and evenly spaced plants for animals, but zooming out to a bigger scale almost always reveals gaps between occupied areas.

In 1976, Eric Charnov (Charnov, 1976) produced a set of important discoveries about patches. He focused on how the patchiness of prey items affects the behavior of a consumer. Essentially, when one prey is found, an experienced predator can form a prediction of how far it must travel, on average, and how long it will take, on average, to find the next prey item. These averages change as the predator exploits a patch; in patches of food that do not replenish themselves, the longer the predator works in the patch the harder it becomes to find food (Fig. 48.1).

The predator is then faced with a critical question: when is it more profitable to stay in a patch, even though returns are diminishing, or when is it more profitable to search out another patch? Experienced predators will know how much work (time and effort) will go into searching out another patch, and how much risk may be involved if unfamiliar terrain needs to be traversed during the search.

Charnov's key prediction is that a predator should leave the patch when the yield rate from that patch dips below the average yield rate for the environment. In other words, an animal getting below-average returns should look for new foraging grounds.

Charnov's paper inspired a generation of ecologists and behaviorists who focused on testing the marginal value theorem and its relevance to real-life

Conceptual Breakthroughs in Ethology and Animal Behavior.
DOI: http://dx.doi.org/10.1016/B978-0-12-809265-1.00048-4

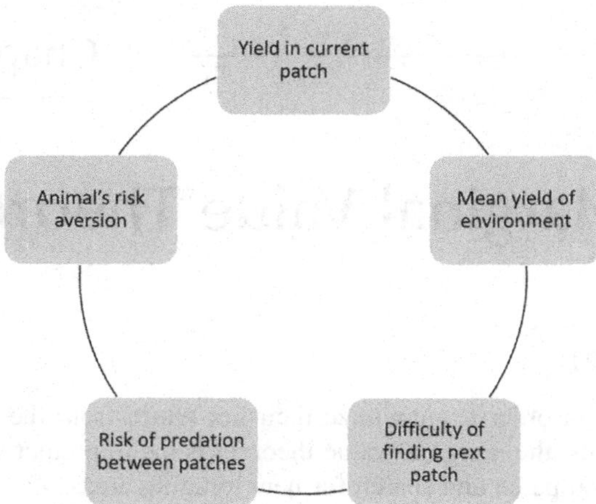

FIGURE 48.1 Information flow in making foraging decisions. Classical marginal value theory predicts that an animal will assess the current yield of the patch in which it is foraging and the mean yield of the environment. If only this information is considered, then the animal should search for a new patch when the yield of the current patch falls below the habitat mean. However, animals tend to be more conservative than this prediction, and this is where other factors are now understood to play key roles. Risk of failure—death due to starvation if a new patch is not found—is not necessarily correlated with mean habitat yield. Knowledge of predation risks between patches can play a key role in decisions. Finally, animals can vary in their risk aversion (see Chapter 77: 2004 Behavioral Syndromes—Personality in Animals), a factor that can cause responses to differ among animals.

predators. One interesting outcome is that animals tend to be a bit more conservative than the theorem dictates; faced with the choice between a certain but slightly less profitable outcome and the uncertainties of searching for a new patch, many animal species wait longer than expected before departing their patch.

This leaves three intriguing dangling threads; the roles of patch replenishment, of the prey animal's experience with predators, and of variation in risk aversion among animals. Nectar is a resource than can be replenished rapidly by a plant. For nectar foragers (its hard to think of them as predators, but the model fits this type of foraging) plants with high replenishment rates allow for continued foraging in a patch even as resource is being removed.

For an inexperienced consumer there is much to be learned about resources and how they are distributed in an environment. It is not surprising that many animals are extremely good at accumulating this type of information even if they have limited general memory capacities. Under the marginal value theorem, experience is critical for foragers in their decision making.

Finally, the assessment of the initial value of a patch has a major influence on subsequent decisions. This is easily observed when watching bees sample flowers; if plants bear multiple flowers and each plant is viewed as a patch, how many flowers does the bee need to sample to make a determination about whether to visit many flowers on that same plant or to move onto the next plant? In most cases the answer is only one or a few; this demonstrates how powerful the application of the marginal value theorem can be in predicting efficient movements by foragers.

IMPACT: 5

The marginal value theorem gave a route for developing hypotheses about how evolution shapes the movement of foraging animals. The marginal value theorem played a major role in shaping optimal foraging theory. Perhaps early work resulting from the marginal value theorem and optimal foraging theory was overly optimistic about the ability of animals to assess yields, remember foraging results, and make calculations of potential future returns, but the revelations about the limitations that animal sensory and cognitive abilities place on optimal foraging have, in themselves, been fascinating.

SEE ALSO

Chapter 55, 1980 The Risk Paradigm; Chapter 57, 1981 Producers and Scroungers; Chapter 63, 1990 Fear; Chapter 77, 2004 Behavioral Syndromes—Personality in Animals.

REFERENCES AND SUGGESTED READING

Brown, J.S., 1988. Patch use as an indicator of habitat preference, predation risk, and competition. Behav. Ecol. Sociobiol. 22, 37–47.

Charnov, E.L., 1976. Optimal foraging, marginal value theorem. Theor. Populat. Biol. 9, 129–136.

McNair, J.N., 1982. Optimal giving-up times and the marginal value theorem. Am. Nat. 119, 511–529.

Pyke, G.H., 1978. Optimal foraging in hummingbirds—testing the marginal value theorem. Am. Zool. 18, 739–752.

Chapter 49

1977 Self-medication

THE CONCEPT

Animals seek out seemingly unusual food items in order to medicate themselves. Natural selection has provided them with the ability to seek out micronutrients as well as compounds that assist in fighting disease and parasites.

THE EXPLANATION

A quiet revolution in how we understand animal diets came with the realization that carnivores do not just eat meat, and that herbivores may not solely rely on plants for their nutrition. Perhaps anyone who owned a dog already had recognized that dogs often eat small amounts of grass and other plants, and many hunters had observed deer licking carcasses of other animals, or eating nestling birds.

These are not oddities or pathological behaviors. Evolution has shaped animals to avoid dietary self-harm, but it has also equipped animals to seek out foods that are not within their normal dietary range. For a herbivore, salt and certain fats are absent or in short supply in their daily diet, and opportunities to gain those nutrients are valuable. Specific appetites for micronutrients and for dietary items that have medicinal value are common in animals.

The example of the dog eating grass suggests a more complicated picture, as dogs often vomit after consuming grass. So is it a mistake for the dog to eat grass, or is it self-medicating? Should a dog owner work to suppress this behavior, or would the dog owner be thwarting the dog's attempt to self-medicate? In addition to possibly working as an emetic, the grass could be a source of dietary fiber for a carnivore that usually does not encounter much fiber in its diet.

Huffman (1997) assembled an overview of this type of self-medication in primates. Until Huffman's publication this had been an overlooked area; scientists neither took the idea that animals might self-medicate seriously, nor had they documented rare and seemingly unusual dietary choices in their field observations. Huffman's work showed that primates often make

Conceptual Breakthroughs in Ethology and Animal Behavior.
DOI: http://dx.doi.org/10.1016/B978-0-12-809265-1.00049-6

FIGURE 49.1 This porcupine may have selected its food based only on carbohydrate, fat, and protein content, but the potential medicinal value of food items should always be considered when studying animal foraging choices. *Photo courtesy Madison Sankovitz.*

interesting dietary choices when they are ill and that those choices can be understood based on the pharmacological potential of the consumed items.

Observations of self-medication led to a fancy word for this field of study: zoopharmacognosy. Since Huffman's key publication, self-medication has been studied in many animals, ranging from bees to lizards to moose, elk, and deer.

Knowing, or suspecting, that an animal has consumed a specific food item in order to self-medicate leads to the question of how the animal "knew" to pursue that food for the specific ailment they were suffering. Animals tend to have very limited dietary ranges in nature because there is so much out there that can be harmful or fatal if eaten. It makes no sense for a sick animal to start randomly sampling a large range of foods, with the motivation that it may find one that is helpful; the odds of doing harm far outweigh the odds of gaining benefit. In fact, animals that self-medicate almost always make a specific choice about what to ingest, without doing a lot of sampling (Fig. 49.1).

So how is this knowledge of self-medication gained? There are two options: first, cultural experience and observations of other animals allows sick animals to already have the information about what foods to choose for self-medication. This is clearly a possibility in highly social species like primates, some rodents such as rats, meerkats, and any other species in which parents "supervise" the dietary choices of their offspring. The second possibility is that natural selection equips an animal with information that causes it to seek out a plant with a specific taste or texture when it has an illness. Examples of this are less well known, but this principle likely applies to female insects that lay their eggs on plants with the potential for protecting their offspring from disease.

From a human perspective, self-medication may be an important and under-recognized part of our biology. Herbs and spices used to flavor foods may have health benefits for consumers, even though the cook for a meal may have no idea about those benefits. Culture and evolution shape dietary preferences to include these beneficial effects.

IMPACT: 4

This concept adds a large dimension to understanding animal diets, food preferences, and sudden proclivities for certain kinds of food that would otherwise be difficult to explain.

REFERENCES AND SUGGESTED READING

Choisy, M., de Roode, J.C., 2014. The ecology and evolution of animal medication: genetically fixed response versus phenotypic plasticity. Am. Nat. 184, S31–S46.

Huffman, M., 1997. Current evidence for self-medication in primates: A multidisciplinary perspective. Yearbook Phys. Anthropol. 104, 171–200.

Krief, S., Hladik, C.M., Haxaire, C., 2005. Ethnomedicinal and bioactive properties of plants ingested by wild chimpanzees in Uganda. J. Ethnopharmacol. 101, 1–15.

Villalba, J.J., Miller, J., Ungar, E.D., Landau, S.Y., Glendenning, J., 2014. Ruminant self-medication against gastrointestinal nematodes: evidence, mechanism, and origins. Parasite 21, 31.

Chapter 50

1977 The Evolution of Mating Systems

THE CONCEPT

A mating system is the social pattern of reproducing animals, such as monogamy or polygyny. Investigators gain insight into the evolution of mating by using the classification of mating systems proposed by Emlen and Oring in 1977.

THE EXPLANATION

A mating system is the pattern of how males and females come together in a population. Surprisingly naturalists and biologists had not developed an overview of patterns of males and females coming together for mating until Emlen and Oring's 1977 paper on mating systems. They identified the main themes of social bonding among animals—monogamy, polygyny, and polyandry.

Monogamy is simple to understand, as it is the pairing of a male and a female. Many bird species are socially monogamous. Variants on monogamy involve surreptitious copulations with other partners, shifting partners from season to season, or abandonment of a partner during a mating season. Monogamy is sometimes, but not always, associated with biparental care.

Polygyny[1] is one male with a group, sometimes called a harem, of females. Males often defend the group of females, dominate the females, or hold resources that are attractive to the females. Emlen and Oring also included leks, assemblages of males that attract females, as a kind of polygyny.

Polyandry is the reverse of polygyny with one female and many males. Resource defense polyandry is known in some species. Social polyandry is much less common than monogamy or polygyny (Parker and Birkhead, 2013).

---

1. Confusingly, the term polygyny is used in the social insect literature to refer to colonies with more than one queen. Monogynous colonies have only one queen. The two sets of terminology are unrelated.

Conceptual Breakthroughs in Ethology and Animal Behavior.
DOI: http://dx.doi.org/10.1016/B978-0-12-809265-1.00050-2

FIGURE 50.1 Distinctive color patterns help animals make mate choices, helping to identify species, sex, and potential quality of mates. *Photo courtesy Madison Sankovitz.*

At the time of Emlen and Oring's work, most biologists thought that the observed social mating system was reflected in actual parent–offspring genetic outcomes. The use of DNA markers and fingerprinting techniques over the next 40 years have shown that the social mating system does not so easily equate to genetic outcomes, as in many species extra-pair copulations overturn the prediction made from social bonds (Griffith et al., 2002). Effectively this gives socially monogamous males and females a route to socially covert polygyny or polyandry (Fig. 50.1).

Emlen and Oring (1977) also made predictions about how mating systems would interact with sexual selection (see Chapter 47: The Handicap Principle). This paper helped to lay the ground for an active and fertile scientific conversation on mate choice that, like the investigations of genetic outcomes of mating systems, extends to the present day. Mate choice then plays into the evolutionary elaboration of characteristics used in choices, such as coloration, body size, and pheromonal representation of immune competency.

The thorough approach used by Emlen and Oring (1977), coupled with a table that laid out in a very clear manner the attributes of each kind of mating system, helped this paper become one of the most cited and influential studies in behavioral ecology.

IMPACT: 9

The importance of this paper peaked as the search for cross-species patterns in animal social behavior came to the fore. The classification proposed by Emlen and Oring is still widely used, but genetic studies have undercut the certainty of knowing mating outcomes from observing behavior alone.

SEE ALSO

Chapter 8, 1859 Darwin and Behavior; Chapter 16, 1947 The Evolution of Clutch Size; Chapter 21, 1954 Life History Phenomena; Chapter 44, 1974 Parent–Offspring Conflict; Chapter 53, 1980 Dispersal; Chapter 60, 1983 Reproductive Skew, Chapter 76, 2002 Social Networks.

REFERENCES AND SUGGESTED READING

Arnqvist, G., Nilsson, T., 2000. The evolution of polyandry: multiple mating and female fitness in insects. Anim. Behav. 60, 145–164.

Clutton-Brock, T.H., 1989. Mammalian mating systems. Proc. Royal Soc. Series B-Biol. Sci. 236, 339–372.

Emlen, S.T., Oring, L.W., 1977. Ecology, sexual selection, and evolution of mating systems. Science 197, 215–223.

Gowaty, P.A., 2013. Adaptively flexible polyandry. Anim. Behav. 86, 877–884.

Griffith, S.C., Owens, I.P.F., Thuman, K.A., 2002. Extra pair paternity in birds: a review of interspecific variation and adaptive function. Mol. Ecol. 11, 2195–2212.

Parker, G.A., Birkhead, T.R., 2013. Polyandry: the history of a revolution. Philosop. Trans. Royal Soc. B-Biol. Sci. 368, 20120335.

West-Eberhard, M.J., 2014. Darwin's forgotten idea: the social essence of sexual selection. Neurosci. Biobehav. Rev. 46, 501–508.

Chapter 51

1978 Animal Models for Depression

THE CONCEPT

This model establishes that an animal exhibits the behavioral manifestations of depression, a pyscho-emotional state previously attributed only to humans.

THE EXPLANATION

Porsolt and colleagues (1978) showed that rats exhibited behavior that matched, in every conceivable way, depression in humans. In their tests, Porsolt et al. (1978) gave rats a hopeless swimming task by placing them in a tank from which there was no escape. Eventually the rats' behavior turned from swimming and attempted escape to something that looks very much like giving up: cessation of activity and immobility.

The next step in the experiments was to treat the rats with pharmacological and physical therapeutics that would be used for depressed humans. The rats became more active (a sign of hopefulness on their part?) when they were treated with antidepressants or with electroconvulsive therapy, but did not respond positively to antianxiety drugs or to tranquilizers.

The realization that an animal could experience something that both behaviorally resembled depression and responded to the same treatments was a critical moment in progress to crediting animals with emotion (see Chapter 74: 2000 Emotion and the Brain) and personality (see Chapter: 2004 Behavioral Syndromes—Personality in Animals). As a sidelight, honeybees show something akin to pessimism if they have negative experiences (Bateson et al., 2011); this could establish a broader generality for emotions like depression in animals. The discovery of a depression-like state in rodents was also an important point in the realization that animal models could be very accurate for testing psycho-pharmaceuticals intended for use on humans.

Conceptual Breakthroughs in Ethology and Animal Behavior.
DOI: http://dx.doi.org/10.1016/B978-0-12-809265-1.00051-4

IMPACT: 7

Although the Porsolt et al. (1978) paper on depression in rats may not have acknowledged the potential of their findings for informing ethology, this publication helped to bridge the gap between ethological work on motivation and drive with the neuroscience of emotion. This paper also helped to break down some of the barriers raised by Morgan's canon (see Chapter 11: 1894 Morgan's Canon) to crediting animals with having emotional lives.

SEE ALSO

Chapter 25, 1960 Motivation and Drive; Chapter 27, 1964 Dopamine and Reward Reinforcement; Chapter 61, 1985 An Animal Model for Anxiety; Chapter 65, 1991 Pain in Animals; Chapter 74, 2000 Emotion and the Brain.

REFERENCES AND SUGGESTED READING

Bateson, M., Desire, S., Gartside, S.E., Wright, G.A., 2011. Agitated honeybees exhibit pessimistic cognitive biases. Current Biol. 21, 1070–1073.

Detke, M.J., Rickels, M., Lucki, I., 1995. Active behaviors in the rat forced swimming test differentially produced by serotonergic and noradrenergic antidepressants. Psychopharmacology 121, 66–72.

Lucki, I., 1997. The forced swimming test as a model for core and component behavioral effects of antidepressant drugs. Behav. Pharmacol. 8, 523–532.

Porsolt, R.D., Anton, G., Blavet, N., Jaffe, M., 1978. Behavioral despair in rats: new model sensitive to antidepressant treatments. European J. Pharmacol. 47, 379–391.

Chapter 52

1978 Theory of Mind

THE CONCEPT

When an animal forms hypotheses about what another animal is thinking, that is a theory of mind. In other words, one animal has a theory about another animal's thought process. This can be an important element of social cognition.

THE EXPLANATION

From a human perspective, it is normal to form hypotheses about the goals and intentions of those around us. We shape our behavior based on what we think others intend to do. We have a *theory of mind* for the intentions and goals of the people around us. Humans engage in multilevel thinking—I act based on how I think the other person thinks I will act. This is an important cognitive process for humans and is a key element of human social behavior.

As with all that the features seemed to set human abilities apart from those of other animals, the barriers to recognizing theory of mind in nonhumans began to break down with the publication of Premack and Woodruff's (1978) work on theory of mind in chimpanzees. Using a single chimpanzee, Sarah, as their subject, Premack and Woodruff (1978) found evidence that she could form theories about the goals and intentions of humans as she observed their behavior.

These experiments led to two decades of wrangling among primatologists, who were divided as to whether chimpanzees only respond to the surface level of prediction about behavior of others, based on knowledge of what they have seen in the past, or if chimpanzees could actually form theories about the goals and intentions of others. Hare et al. (2000) provide perhaps the most conclusive evidence for theory of mind in chimpanzees. Call and Tomasello (2008) review the history of the numerous experiments designed to test Premack and Woodruff's original observation. They conclude that chimpanzees do likely have a theory of mind, but interestingly, that chimpanzees cannot identify false beliefs held by other individuals.

Conceptual Breakthroughs in Ethology and Animal Behavior.
DOI: http://dx.doi.org/10.1016/B978-0-12-809265-1.00052-6

Scrub jays show similar abilities to assess the goals of other birds in their population (Clayton et al., 2007). Because testing for a theory of mind in nonhumans involves a large degree of inference about thought processes in a nonspeaking animal, this is likely to always be a controversial area of inquiry.

IMPACT: 4

The big questions about theory of mind are whether we can really test for its existence and how broadly distributed this type of cognition might be in the animal world. The jury is very much still out on both of these questions. At present, theory of mind is an interesting topic that is actively being researched in the science of cognition.

SEE ALSO

Chapter 37, 1973 Episodic Memory.

REFERENCES AND SUGGESTED READING

Call, J., Tomasello, M., 2008. Does the chimpanzee have a theory of mind? 30 years later. Trends Cogn. Sci. 12, 187–192.

Clayton, N.S., Dally, J.M., Emery, N.J., 2007. Social cognition by food-caching corvids. The western scrub-jay as a natural psychologist. Philos. Trans. Royal Soc. B: Biol. Sci. 362, 507–522.

Hare, B., Call, J., Agnetta, B., Tomasello, M., 2000. Chimpanzees know what conspecifics do and do not see. Anim. Behav. 59, 771–785.

Kaminski, J., Emery, N.J., 2010. Social cognition and theory of mind. In: Breed, M.D., Moore, J. (Eds.), The Encyclopedia of Animal Behavior. Elsevier Ltd, Oxford.

Premack, D., Woodruff, G., 1978. Does the chimpanzee have a theory of mind? Behav. Brain Sci. 1, 515–526.

Whiten, A., 2013. Humans are not alone in computing how others see the world. Anim. Behav. 86, 213–221.

Chapter 53

1980 Dispersal

THE CONCEPT

Young animals leave their parents' habitat in search of unoccupied habitat, more productive habitat, or mating opportunities with unrelated animals. The age and sex of dispersing animals is predictable based on mating system and life history characteristics.

THE EXPLANATION

Greenwood's (1980) exploration of how dispersal interacts with animals' mating systems came soon after Emlen and Oring's (1977) exposition on mating systems. In birds and mammals, young animals may be much impacted by their parents' continued presence and use of resources. Competition with parents is potentially a losing prospect, but dispersing is highly risky as young animals may have to traverse inhospitable terrain, be exposed to predators, and may find that the habitat where they arrive at is equally crowded. Observations leading up to Greenwood's paper had noted that male young sometimes dispersed while females sometimes dispersed, but discovering patterns had proven difficult.

Greenwood (1980) built on the accumulation of natural history knowledge, noting that female birds were more often dispersers, while in mammals it was the males that more often dispersed. Staying close to home, or philopatry, appears in his analysis to be predicted by the mating system. Mammals are more often polygynous, so that males move away from a group that includes their mother and sisters. Birds are more often monogamous, perhaps favoring female movement because the future reproduction of their father depends on retention of his territory.

Dispersal is important in finding new habitat for feeding and nesting, but it comes with risks. Dispersing animals may be more exposed to predation or starvation. Game theory (see Chapter 38: 1973 Game Theory) helps to analyze situations such as this, in which animals of each sex could attempt to outwait the other.

A contrary view about mammalian dispersal notes that in some mating systems (see Chapter 50: 1977 The Evolution of Mating Systems) females may find it substantially easier to join new groups than males, thus making female

Conceptual Breakthroughs in Ethology and Animal Behavior.
DOI: http://dx.doi.org/10.1016/B978-0-12-809265-1.00053-8

dispersal less risky. Wild horses and a variety of primates possibly exemplify this point.

Dispersal is also extremely important in helping animals to obtain an appropriate level of outbreeding. Animal in populations that do not disperse at all risk becoming inbred and suffering from the effects of lost genetic variation. Under circumstances in which dispersal is limited by geography, such as island populations, the impacts of mate choice and sexual selection may be muted, as there can be a high degree of similarity among potential mates. As with finding new habitat in which to live, finding new genetic terrain is a complicated proposition. This is a difficult cost–benefit analysis that takes generations of evolution to solve.

Greenwood's (1980) paper is among the most highly cited papers in the fields of animal behavior, ethology, and behavioral ecology. This is partly because of his thorough data review and insightful analysis. But not everyone agrees with his fundamental point that the male/female difference in dispersal between mammals and birds is a derivative of their mating systems. Also, the natural world is full of exceptions and more knowledge of individual species' dispersal patterns, have not undone the generalization, but they have helped us to understand that a species' evolutionary history can drive dispersal patterns that run counter to the grain of the generality.

IMPACT: 5

Given the importance of dispersal in animal's lives, it is very surprising that stronger consideration had not been given, prior to Greenwood's 1980 publication, to patterns of dispersal and the evolutionary forces that might drive male–female differences in dispersal. This led to a broader consideration of contrasts between birds and mammals and to the interaction of life history characteristics with dispersal patterns.

SEE ALSO

Chapter 21, 1954 Life History Phenomena; Chapter 44, 1974 Parent–Offspring Conflict; Chapter 50, 1977 The Evolution of Mating Systems.

REFERENCES AND SUGGESTED READING

Dobson, F.S., 2013. The enduring question of sex-biased dispersal: Paul J. Greenwood's (1980) seminal contribution. Anim. Behav. 85, 299–304.

Emlen, S.T., Oring, L.W., 1977. Ecology, sexual selection, and evolution of mating systems. Science 197, 215–223.

Greenwood, P.J., 1980. Mating systems, philopatry and dispersal in birds and mammals. Anim. Behav. 28, 1140–1162.

Greenwood, P.J., Harvey, P.H., 1982. The natal and breeding dispersal of birds. Ann. Rev. Ecol. Syst. 13, 1–21.

Johnson, M.L., Gaines, M.S., 1990. Evolution of dispersal-theoretical-models and empirical tests using birds and mammals. Ann. Rev. Ecol. Syst. 21, 449–480.

Chapter 54

1980 Semantic Communication

THE CONCEPT

Animals have the flexibility to combine signals in patterns that carry specific meanings. Rearranging the signals changes the meaning, and regional dialects can form based on cultural traditions in signaling.

THE EXPLANATION

Human language is based on several central concepts, including semantics and syntax. Semantic communication is when a signal refers specifically to an object or stimulus in the environment. For human language semantic communication means the use of words. Seyfarth et al.'s (1980a) study on vervet monkey alarm calls established, for the first time, that a nonhuman could use semantic communication.

Seyfarth et al. (1980b) found the vervet monkeys in their study population could specify by vocalization which threat (hawk, snake, and so on) was approaching. Monkeys that heard the threat-specific calls then responded appropriately to move away from the specific threat. While the conclusion that these monkey calls are semantic has generated controversy and the use of a concept that describes human language in interpreting animal communication could amount to anthropomorphism, this study has impelled considerable interesting studies in other primates and a general consideration of how to define language in a way that encompasses animals as well as humans.

The second concept underlying language, syntax, is the ordering of signals in a chain in a way that supports the collective meaning of the signals. In English a very simple syntax for a sentence is subject:verb:object. While humans can sometimes puzzle out the meaning of a series of words that are in nonsensical order, syntax establishes sense by ordering of words. Scientists have argued that chimpanzees and other apes that can use sign language or electronic boards to communicate with their human handlers are using syntax as well as semantics (Hauser et al., 2002).

Conversely, the fabric of the argument for nonhuman use of syntax and semantics is stretched very thin when applied to honeybees, which use

Conceptual Breakthroughs in Ethology and Animal Behavior.
DOI: http://dx.doi.org/10.1016/B978-0-12-809265-1.00054-X

symbolic representations in their dances to convey the distance, direction and value of food resources. Perhaps the biggest difference between bees and primates is that the information for bees is encoded genetically, while the linguistic information used by at least some primates is learned and influenced by culture.

IMPACT: 2

Semantic communication has been an interesting and important facet of the overall progress in research on animal communication. While not applicable to a very broad range of animals, special interest is gained from looking at monkeys and apes as stepping-stones to human linguistic abilities.

SEE ALSO

Chapter 37, 1973 Episodic Memory; Chapter 52, 1978 Theory of Mind; Chapter 67, 1992 Working Memory.

REFERENCES AND SUGGESTED READING

Fichtel, C., Kappeler, P.M., 2002. Anti-predator behavior of group-living Malagasy primates: mixed evidence for a referential alarm call system. Behav. Ecol. Sociobiol. 51, 262–275.

Hauser, M.D., Chomsky, N., Fitch, W.T., 2002. The faculty of language: what is it, who has it, and how did it evolve? Science 298, 1569–1579.

Manser, M.B., 2013. Semantic communication in vervet monkeys and other animals. Anim. Behav. 86, 491–496.

Seyfarth, R.M., Cheney, D.L., Marler, P., 1980a. Monkey responses to 3 different alarm calls: evidence of predator classification and semantic communication. Science 210, 801–803.

Seyfarth, R.M., Cheney, D.L., Marler, P., 1980b. Vervet monkey alarm calls: semantic communication in a free-ranging primate. Anim. Behav. 28, 1070–1094.

Chapter 55

1980 The Risk Paradigm

THE CONCEPT

Animals are sensitive to the risks involved in their actions and the assessment of the potential costs and benefits of behavior includes the effects of risk. In the risk paradigm, the variability of the environment for foraging rewards is more important than the average foraging reward. Highly variable yields from foraging are inherently more risky, as they can include rewards that will be below starvation level, and foragers may avoid risk if they are already in marginal survival situations.

THE EXPLANATION

Caraco et al. (1980) revolutionized foraging research by introducing risk sensitivity. Studying juncos, they found that the birds responded not only to the amount of food available in tests—the energetic reward from foraging—but also to the variability of the food reward. This was a particularly important insight as up to this point investigations of foraging behavior had generally assumed that animals respond to the average yield of the environment, and risks imposed by variation were not well incorporated into models. Caraco et al. (1980) employed a utility function to predict that an animal's risk aversion would predictably change according to its current energy budget (the "budget rule").

Risk sensitivity can occur on a cognitive level, based on direct experience or the observation of other animals. Risk can also act as a selective force, shaping behavior through incorporation of the outcomes of risky behavior into genetic information that underlies choices. If risk sensitivity is cognitive, then quite a lot is being asked of the animal in terms of its ability to monitor its own status, to assess the variability of yields from the environment, and then to make calculated moves based on the budget rule. Similarly, evolved sensitivity still would require comprehensive assessment of foraging returns from the environment.

Kacelnik and Moudon (2013) critique the risk paradigm, pointing out that relatively few studies support the utility function proposed by Caraco et al. (1980). Perhaps this is because such calculations are beyond the capacity of

Conceptual Breakthroughs in Ethology and Animal Behavior.
DOI: http://dx.doi.org/10.1016/B978-0-12-809265-1.00055-1

most animals to assess their environment in such a comprehensive way. Alternatively, desperation may actually yield more risky behavior, thus confounding the budget rule. Nevertheless, Kacelnik and Moudon (2013) acknowledge that the risk paradigm model propelled two decades of interesting and productive research.

Kacelnik and Moudon (2013) point out that: "in economics. . .utility functions are mostly assumed to be stable for individuals and homogeneous across populations." In other words, at least in these authors' view, human economic science has assumed that people do not adjust their risk aversion according to the variability of the economic climate. This is an interesting counterpoint to the fact that many animal species are known to express behavior along a shyness to boldness gradient, which can vary within an individual over the course of its lifetime or among individuals within a population (see Chapter 77: 2004 Behavioral Syndromes—Personality in Animals). Integration of the behavioral syndrome concept with the risk paradigm could, indeed, be a fruitful path for further investigation.

IMPACT: 2

The risk paradigm remodeled how we view the behavior of foraging animals. Risk sensitivity explains deviations from the predictions of optimal foraging theory (see chapter: 1976 Marginal Value Theorem). The paradigm could apply in other contexts in which animals are faced with variable environments. Even though acceptance of the paradigm is open to question, its impact on research has been extensive.

SEE ALSO

Chapter 48, 1976 Marginal Value Theorem; Chapter 63, 1990 Fear; Chapter 77, 2004 Behavioral Syndromes—Personality in Animals.

REFERENCES AND SUGGESTED READING

Caraco, T., Martindale, S., Whittam, T.S., 1980. An empirical demonstration of risk-sensitive foraging preferences. Anim. Behav. 28, 820–830.

Kacelnik, A., Mouden, C.E., 2013. Triumphs and trials of the risk paradigm. Anim. Behav. 86, 1117–1129.

Chapter 56

1981 Prisoner's Dilemma

THE CONCEPT

The prisoner's dilemma is a simple game in which players have the choice of collaborating with another prisoner or defecting and helping the prison guards. This game allows the exploration of conditions in which animals might behave altruistically by collaborating with others.

THE EXPLANATION

How does cooperation work out in the daily lives of animals? Axelrod and Hamilton (1981) used a simple game to illustrate how conditions might enforce reciprocity between pairs of animals. The prisoner's dilemma game, which is studied in economics, sociology, and psychology, as well as biology, serves as a perfect platform for developing models of the concept of cooperation.

The basic outline of the prisoner's dilemma is this: two prisoners are given the opportunity to either cooperate with the each other, by not giving information to the guards, or to defect, giving the guards incriminating information about the other prisoner. Both prisoners are punished a little when they refuse to yield information. If one defects and the other doesn't, then the defector is rewarded and the nondefector gets a major punishment. If both defect, both receive major punishments. The prisoners take turns, each at their turn deciding whether to cooperate with the other prisoner or defect. Thinking about this situation for a minute, one route is to remain silent and receive a series of minimal punishments.

But what happens when the other prisoner defects and you receive the major punishment? On your next turn you can punish the other prisoner back by defecting. This strategy, of cooperating and then punishing the other prisoner if they defect, is called tit-for-tat. Generally speaking, tit-for-tat is a winning strategy if the ending time for the game is not known. A strategy that wins and cannot be dislodged by another strategy is an Evolutionarily Stable Strategy (ESS) (see Chapter 38: 1973 Game Theory). In games of prisoner's dilemma with fixed time limits, knowing that you will have the last turn is a huge advantage, as you can always defect on the last turn, receive the reward, and go unpunished by the other prisoner (Fig. 56.1).

Conceptual Breakthroughs in Ethology and Animal Behavior.
DOI: http://dx.doi.org/10.1016/B978-0-12-809265-1.00056-3

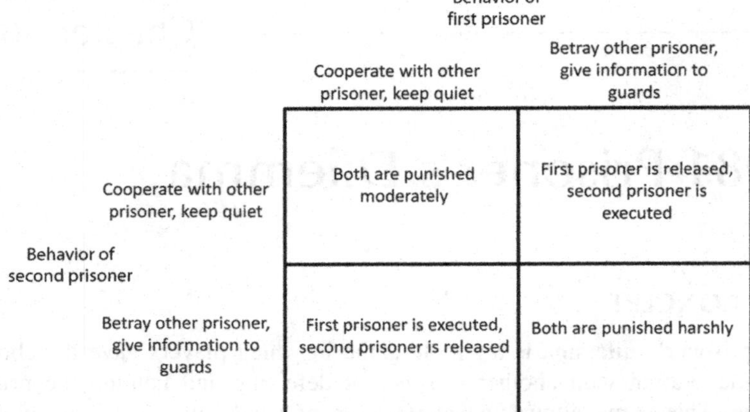

FIGURE 56.1 An example of a payoff matrix for prisoner's dilemma. In this version, the extremity of one of the punishments, execution, precludes playing the tit-for-tat strategy. Obviously if either prisoner knows that they have the last move, they should defect. If neither player knows the length of the game, this particular matrix favors extended cooperation. Less severe punishments open the door for tit-for-tat strategies. One of intriguing aspects of prisoner's dilemma is that the payoff matrix can be manipulated to test for levels of risk-aversion on the part of the players.

IMPACT: 5

The prisoner's dilemma is seemingly simple-minded but in reality it is almost infinitely complex. Starting with the prisoner's dilemma allows deep exploration of the evolution, culture, economics, and sociology of cooperation in all animals.

SEE ALSO

Chapter 28, 1964 Inclusive Fitness and the Evolution of Altruism; Chapter 35, 1971 Reciprocal Altruism; Chapter 36, Selfish Herds; Chapter 38, 1973 Game Theory; Chapter 45, 1975 Group Selection.

REFERENCES AND SUGGESTED READING

Axelrod, R., 1984. The Evolution of Cooperation. Basic Books, New York, NY, 223 pp.

Axelrod, R., Hamilton, W.D., 1981. The evolution of cooperation. Science 211, 1390–1396.

Hammerstein, P., Noe, R., 2016. Biological trade and markets. Philos. Trans. Royal Soc. B-Biol. Sci. 371, 20150101.

Noe, R., 2006. Cooperation experiments: coordination through communication versus acting apart together. Anim. Behav. 71, 1–18.

Chapter 57

1981 Producers and Scroungers

THE CONCEPT

In social foraging groups some animals are producers; they discover food and find new resources. Others are scroungers that exploit the findings of producers. Scroungers can be either the same species as the producers, as in a pride of lions, or a different species, as in a mixed-species flock of birds. Scrounging is a cost of social behavior.

THE EXPLANATION

Food robbing is a very common behavior among animals. Casual observations of gulls at the ocean shore, groups of crows, or of dogs show that attempts to steal food may be more common than actual discoveries of food. Among gulls a food item may be repeatedly stolen, passing from bird to bird, until it is finally consumed. Barnard and Sibly (1981) called the animals that take already-discovered food scroungers.

On the other hand, some animals search out new find items and new foraging territories. These are producers (Barnard and Sibly, 1981). While time and effort go into both scrounging and producing, producing is the only route by which new nutrients can enter the social group.

Barnard and Sibly's (1981) point is elegantly simple. Experiencing scrounging is a cost, for the producers, of being in a social group. This argument then leads to the question of why an animal is in a social group when it must pay a cost for membership. One hypothesis is that there are counterbalancing benefits of sociality such as protection from predators (see Chapter 36: Selfish Herds, and Chapter 39: 1973 Many Eyes Hypothesis). Alternatively, it may cost more to avoid scroungers or to drive them away than it does to tolerate their presence.

The producer–scrounger model lends itself to game theory analysis (see Chapter 38: 1973 Game theory) and is an excellent counterpoint to models that require kinship to explain social behavior (see Chapter 28: 1964 Inclusive Fitness and the Evolution of Altruism).

Conceptual Breakthroughs in Ethology and Animal Behavior.
DOI: http://dx.doi.org/10.1016/B978-0-12-809265-1.00057-5

IMPACT: 2

Barnard and Sibly's (1981) model gained some attention and stimulated further discussion about how to incorporate the exploitative role of scroungers into social evolution theory (e.g., Vickery et al., 1991). The model is well-worth revisiting at this point with three decades of perspective on social evolution from both modeling efforts and empirical data collection.

SEE ALSO

Chapter 28, 1964 Inclusive Fitness and the Evolution of Altruism; Chapter 36, Selfish Herds; Chapter 38, 1973 Game Theory; Chapter 39, 1973 Many Eyes Hypothesis.

REFERENCES AND SUGGESTED READING

Barnard, C.J., Sibly, R.M., 1981. Producers and scroungers: a general-model and its application to captive flocks of house sparrows. Anim. Behav. 29, 543–550.

Heinsohn, R., Packer, C., 1995. Complex cooperative strategies in group-territorial African lions. Science 269, 1260–1262.

Vickery, W.L., Giraldeau, L.A., Templeton, J.J., Kramer, D.L., Chapman, C.A., 1991. Producers, scroungers, and group foraging. Am. Nat. 137, 847–863.

Chapter 58

1982 The Hamilton–Zuk Hypothesis

THE CONCEPT

The ability to produce signals to attract mates may be affected by parasite and disease loads; animals with better immune functions produce better signals. This can make sexually selected signals, such as plumage color in birds, honest reflections of a potential mates immunological capacity.

THE EXPLANATION

Suppose that females of a species all choose to select their mates for a special trait, such as the color of their feathers. It should be obvious, from what we know about natural selection, that it will not take many generations for all males to share the same color. The bright reds of male northern cardinals and the heads of male house finches are good examples of this kind of outcome from natural selection. Ironically, whatever information the red color gave females at the beginning stages of the selective process may have been lost over evolutionary time. This exemplifies sexual selection, one of the selective processes affecting behavior that Darwin identified. If you need a quick refresher on sexual selection, you might want to look back at Chapter 8, 1859 Darwin and Behavior.

But you would expect selection on the females to favor choosing male attributes that are inescapably linked to valuable information about the male. One way this happens is if the male's feature is costly to produce. Zahavi explored costly signals in the development of his handicap principle (see Chapter 47: The Handicap Principle).

W.D. Hamilton and M. Zuk took another angle in looking at this question. What if disease or parasitism reduces a male's ability to produce a signal? Specifically, what if having parasites makes it more difficult for male birds to produce bright plumage colors? Essentially, the argument is that males with effective immune systems have better health due to fewer parasites and consequently will be able to produce bright feathers. The

Conceptual Breakthroughs in Ethology and Animal Behavior.
DOI: http://dx.doi.org/10.1016/B978-0-12-809265-1.00058-7

FIGURE 58.1 According to the Hamilton–Zuk hypothesis, healthier animals are able to produce brighter colors, making coloration an honest signal of health and immune competency. *Photo courtesy Madison Sankovitz.*

Hamilton–Zuk hypothesis is deceptively simple to describe, given the difficulties that generations of biologists have encountered in testing it.

Imagine, in the wake of Hamilton and Zuk's (1982) publication, scientists avidly chasing birds trying to collect fresh feces for parasitological examination. Other scientists focused on external parasites, such as mites, and still others collected blood samples to investigate parasites like malaria (yes, birds do get their own forms of malaria). This cavalcade of investigation made the Hamilton–Zuk hypothesis one of the most studied questions in animal behavior in the 1980s and 1990s (Fig. 58.1).

Despite dozens, perhaps hundreds, of studies that address the Hamilton–Zuk hypothesis, the jury is still out on whether data broadly supports the idea. The most important outcome of Hamilton and Zuks' contribution has been a thorough look at variability in mate choice signals and careful considerations of how the quality of an animal's immune system can be reflected in features like plumage color and the symmetry of an animal. Mate choice is one of the most important and complex acts in an animal's life, and studies of mate choice reveal the complexity, rather than discovering simple paradigms that govern all animals.

IMPACT: 8

The Hamilton–Zuk hypothesis shaped much of the conversation about mate choice and sexual selection over the two decades following its publication. While there have been ups and downs in the degree to which the hypothesis has been accepted, the impetus it gave for research and data collection has been tremendous.

SEE ALSO

Chapter 47, The Handicap Principle; Chapter 50, 1977 The Evolution of Mating Systems.

REFERENCES AND SUGGESTED READING

Candolin, U., 2003. The use of multiple cues in mate choice. Biol. Rev. 78, 575–595.

Getty, T., 2002. Signaling health versus parasites. Am. Nat. 159, 363–371.

Hamilton, W.D., Zuk, M., 1982. Heritable true fitness and bright birds: a role for parasites? Science 218, 384–387.

Martin, J., Lopez, P., 2015. Condition-dependent chemosignals in reproductive behavior of lizards. Horm. Behav. 68, 14–24.

Wedekind, C., 1992. Detailed information about parasites revealed by sexual ornamentation. Proc. Royal Soc. B-Biol. Sci. 247, 169–174.

Westneat, D.F., Birkhead, T.R., 1998. Alternative hypotheses linking the immune system and mate choice for good genes. Proc. Royal Soc. B-Biol. Sci. 265, 1065–1073.

Chapter 59

1982 The Hippocampus and Navigation

THE CONCEPT

The Morris water maze is a useful technique for studying orientation and spatial learning in rodents. It combines an aversive stimulus, immersion in water, with escape routes that can be placed to test the animal's use of previous experiences in shaping its movements. Its earliest application was in the establishment of the hippocampus as the central integrator in vertebrate navigation.

THE EXPLANATION

The brain is not just a large black box that receives information and dispenses instructions. It is anatomically layered, multifaceted, and exceedingly complex. The vertebrate brain was built, over the course of hundreds of millions of years of evolution, by starting with the relatively simple structure found in fish and adding onto that as evolution required more complex functions. The brains of birds and mammals are structurally very much like old houses that have been remodeled and had room additions made many times; they lack a simple, clear and direct functional plan, they often contain parts with redundant or overlapping functions, and the soundness of their function sometimes depends on surprising interconnections between parts.

Because of all of this, it can be very difficult to assign a specific function to a specific area of the brain. Historically, the major technique for discovering what parts of the brain are essential for any given function is to create lesions and then to observe what functions are lost because of those lesions. Functional magnetic resonance imaging has, in the last two decades, added a new tool for these sorts of studies as it enables measurement of brain activity while an animal is performing an activity. Nevertheless, pinning down function to anatomy is extremely difficult.

The study by Morris and colleagues (1982) is regarded as particularly significant because they were able to assign, with a high level of certainty,

Conceptual Breakthroughs in Ethology and Animal Behavior.
DOI: http://dx.doi.org/10.1016/B978-0-12-809265-1.00059-9

place navigation to the hippocampus. The hippocampus is one of the evolutionarily "older" parts of the brain, and this finding supports the idea that navigation is a problem held in common by all animals and that in the course of evolution the location for this function was conserved in the vertebrate brain. The Morris et al. (1982) publication led the way to a deep understanding of the neurobiology of navigation.

In addition to the fundamental finding about anatomy and function in the brain, the Morris water maze (Morris, 1984) became a widely accepted research tool. It involves placing a rat in a tub of water with high walls so it cannot escape by climbing up the edge of the enclosure. Generally the tub is circular, so no cues are given to the rat by the shape of the enclosure's boundaries, overhead lighting, if present, is directly over the center of the enclosure, and olfactory cues are equally well controlled. The rat can find a platform to stand on by swimming around the enclosure. In most experiments the platform is slightly below or at the surface of the water, so the rat cannot see the platform.

In many ways this setup is analogous to a Skinner box (see Chapter 13: 1938 Skinner and Learning), which is used to test hypotheses regarding operant conditioning. In the case of the Morris water maze, the object is to test hypotheses regarding learned navigational routes and spatial memory. The aversive nature of the swimming condition is paired with the reward of finding the platform. While Morris' original experiments employed rats as experimental subjects, the technique has also been used with mice and could potentially be applied to a broad range of animals.

Morris' original work (1982) focused on spatial learning. The most common applications of the technique, from the beginning of its use, have been to assess the effects of neural lesions, neuropharmacological interventions, and neurogenetic modifications on spatial memory. The Morris water maze gives investigators a tool to determine the effect of changes in the brain on specific aspects of spatial learning performance. Navigational functions can then be assigned to specific regions of the brain and hold the potential for developing interventions in disorders that affect human navigational abilities.

In recent years, doubts have been raised about the ethics of using the potential for drowning in an inescapable body of water as a training device. Experience shows that rats quickly learn that the apparatus is a problem to be solved, and rats, depending on their temperament (see the watermaze.org website) and genetics, show little sign of panic in the apparatus. Experimental protocols require that rats not be allowed to swim too long before being guided to the platform. Much knowledge about learning has been gained from using the apparatus, but a healthy and open conversation about the welfare of the animals being tested is quite important.

IMPACT: 9

The assignment of specific functions to anatomical regions in the brain is challenging and the finding that spatial learning and navigation reside in the hippocampus was a key development in the quest to explore where cognitive functions lie in the brain. Next to the Skinner box, this is one of the most commonly used techniques in the experimental psychology of learning. The Morris water maze has yielded highly important insights into the anatomy, chemistry, and genetics of the brain.

SEE ALSO

Chapter 17, 1948 Cognitive Maps.

REFERENCES AND SUGGESTED READING

Baker, R., 2013. Water Maze. http://www.watermaze.org/.

Burgess, N., Maguire, E.A., O'Keefe, J., 2002. The human hippocampus and spatial and episodic memory. Neuron 35, 625–641.

D'Hooge, R., De Deyn, P.P., 2001. Applications of the Morris water maze in the study of learning and memory. Brain Res. Rev. 36, 60–90.

Maguire, E.A., Gadian, D.G., Johnsrude, I.S., Good, C.D., Ashburner, J., Frackowiak, R.S.J., et al., 2000. Navigation-related structural change in the hippocampi of taxi drivers. Proc. Natl. Acad. Sci. USA 97, 4398–4403.

Morris, R., 1984. Developments of a water-maze procedure for studying spatial-learning in the rat. J. Neurosci. Methods 11, 47–60.

Morris, R.G.M., Garrud, P., Rawlins, J.N.P., O'Keefe, J., 1982. Place navigation impaired in rats with hippocampal-lesions. Nature 297, 681–683.

Richard, R.G.M., 2008. The Morris water maze. Scholarpedia 3, 6315, http://scholarpedia.org/article/Morris_water_maze.

Chapter 60

1983 Reproductive Skew

THE CONCEPT

Reproductive skew is the variation in reproduction among society members. According to skew theory, this variance can be explained by how social dominance interacts with the options that subordinates have for leaving the group or resisting the dominant members of the group.

THE EXPLANATION

Reproductive skew is Vehrencamp's (1983) model for how access to reproduction is distributed among members of animal societies. When animals live in groups, life is rarely completely egalitarian. Unequal access to resources, dominance behavior, and social status gained from parents are all factors that contribute to animals within a society having very different lives. These differences are often reflected in reproduction, with some group members contributing more offspring to the next generation than others. This variance is called reproductive skew.

Skew models differ from kin selection (see Chapter 28: 1964 Inclusive Fitness and the Evolution of Altruism) in that the interests of the animals in the group are considered to be in conflict. In contrast, kin selection explanations for reproductive variance in groups posit that group members act to maximize their inclusive fitness.

In transactional models for reproductive skew, dominants may concede reproduction to subordinates in order to gain their cooperation, or subordinates may restrain their reproduction in order to ensure continued membership in the group (Buston et al., 2007). As in many models of animal behavior, these decisions about concession and restraint could conceivably be made on a cognitive level with animals assessing their prospects in real time and making concessions or accepting restraints based on current information. Alternatively, natural selection over generations can shape such decisions. The natural selection route makes it possible to apply skew models to animals that lack the cognitive abilities to make real time assessments.

Conceptual Breakthroughs in Ethology and Animal Behavior.
DOI: http://dx.doi.org/10.1016/B978-0-12-809265-1.00060-5

Tug of war models for skew are more straightforward, as animal's competitive abilities determine the outcomes of contests for access to reproduction. More sophisticated models take into account both transactional and tug of war components.

Cooperative breeding, which is when brood care is shared among animals in a social group, is the most common context to which skew theory has been applied. Reproductive skew analysis helps investigators to understand gains and losses in the balance between competition and cooperation in reproduction.

IMPACT: 3

Skew models have not gained the prominence held by kin selection and group selection models for explaining variance in reproduction within social groups, but they have been important in studies of cooperatively breeding birds and mammals. The most important outcome of skew theory has been a thorough consideration of the benefits of dominance and the costs of being subordinate.

SEE ALSO

Chapter 28, 1964 Inclusive Fitness and the Evolution of Altruism; Chapter 35, 1971 Reciprocal Altruism; Chapter 45, 1975 Group Selection.

REFERENCES AND SUGGESTED READING

Buston, P.M., Reeve, H.K., Cant, M.A., Vehrencamp, S.L., Emlen, S.T., 2007. Reproductive skew and the evolution of group dissolution tactics: a synthesis of concession and restraint models. Anim. Behav. 74, 1643–1654.

Emlen, S.T., 1995. An evolutionary theory of the family. Proc. Natl. Acad. Sci. USA 92, 8092–8099.

Ratnieks, F.L.W., Foster, K.R., Wenseleers, T., 2006. Conflict resolution in insect societies. Ann. Rev. Entomol. 51, 581–608.

Reeve, H.K., Sheng-Feng, S., 2013. Unity and disunity in the search for a unified reproductive skew theory. Anim. Behav. 85, 1137–1144.

Vehrencamp, S.L., 1983. A model for the evolution of despotic versus egalitarian societies. Anim. Behav. 31, 667–682.

Vehrencamp, S.L., 2000. Evolutionary routes to joint-female nesting in birds. Behav. Ecol. 11, 334–344.

Chapter 61

1985 An Animal Model for Anxiety

THE CONCEPT

Animal models for human psychiatric conditions have two important functions. First, they allow for trials of behavioral treatments and medications without the use of human subjects. Second, they help to establish the generality of emotion and personality beyond being concepts that are only applied to humans. Pellow et al.'s (1985) study of rodents established the validity of an assay for anxiety.

THE EXPLANATION

For many animals, unsheltered locations are also unsafe locations. Lack of shelter can lead to exposure to predators as well as to unfavorable environmental conditions. A long tradition of behavioral assays of relies on Open-Field Testing (OFT) (Hall and Ballachey, 1932). If a wall is present in an arena, animals tend to cling to that wall when making explorations, particularly when they are new to the arena. Excursions from the wall are akin to movement away from shelter. In OFT, there is no wall; animals move in a seemingly boundless arena. While some controversy has brewed over the validity of OFT for measuring anxiety and emotionality, OFT has been used in numerous studies to assess anxiety responses in animals and to compare anxiety responses between sexes, across ages, between genetic strains, or among species.

Pellow et al. (1985) used a variant of OFT in which animals were placed in either open or closed arms in a maze. As with OFT, the open arms may represent a less safe condition in the animal's perception, and indeed more anxiety-related behaviors were observed in animals in the open arms.

Pellow et al. (1985) extended their findings from behavioral to pharmacological, treating subject animals with chemicals known to affect anxiety or depression, or to have tranquilizing effects. Their critically important finding was that the compounds that impact anxiety in humans changed the behavior

Conceptual Breakthroughs in Ethology and Animal Behavior.
DOI: http://dx.doi.org/10.1016/B978-0-12-809265-1.00061-7

of the animals in the open arms, while antidepressants and tranquilizers did not. This was taken to support the validity of the assay for anxiety, as well as to establish that the neurochemical architecture that underlies anxiety is similar in rodents and humans.

This example of establishing an animal model for a subjective behavioral state—anxiety—follows on the conceptual development of an animal model for depression (see Chapter 51: 1978 Animal Models for Depression). Establishing an animal model for an emotional state requires creating an operational definition for the state. In other words, the emotional state must be defined in a way that allows measurement of either behavioral expression or a neural correlate of the state. As with depression, having an animal model for anxiety requires acceptance that nonhuman animals can be anxious.

IMPACT: 4

Animal models for the neurochemical bases of personality and emotion are extremely important. This foundational study helped to establish the generality, at least among mammals, of functions once thought to be restricted to primates. This line of research, which derives from comparative psychology, has great potential for integration with research on behavioral syndromes (see Chapter 77: 2004 Behavioral Syndromes—Personality in Animals).

SEE ALSO

Chapter 25, 1960 Motivation and Drive; Chapter 27, 1964 Dopamine and Reward Reinforcement; Chapter 51, 1978 Animal Models for Depression; Chapter 65, 1991 Pain in Animals.

REFERENCES AND SUGGESTED READING

Denenberg, V.H., 1969. Open field behavior in the rat: what does it mean? Ann. NY Acad. Sci. 159, 852–859.

Gould T.D., Dao D.T., Kovacsics C.E., 2009. The open field test. In Mood and Anxiety Related Phenotypes in Mice. Volume 42 of the series Neuromethods, pp. 1–20.

Hall, C.S., Ballachey, E.L., 1932. A study of the rat's behavior in a field: a contribution to method in comparative psychology. Univer. California Publicat. Psychol. 6, 1–12.

Miczek, K.A., Weerts, E.M., Vivian, J.A., Barros, H.M., 1995. Aggression, anxiety and vocalizations in animals—GABA(A) and 5-HT anxiolytics. Psychopharmacology 121, 38–56.

Pellow, S., Chopin, P., File, S.E., Briley, M., 1985. Validation of open-closed arm entries in an elevated plus-maze as a measure of anxiety in the rat. J. Neurosci. Methods 14, 149–167.

Chapter 62

1988 Brood Parasitism

THE CONCEPT

Cuckoos lay their eggs in the nests of other birds, relying on those species to rear their young. Given the time and effort that it takes to rear a chick, the expectation would be for the host to evolve ways to detect and expel the cuckoo's egg. The cuckoo would then evolve new deceptions to ensure acceptance of its eggs. In this classic paper, Davies and Brooke (1988) consider the strategies and counterstrategies used by cuckoos and wood warblers in this evolutionary tug of war.

THE EXPLANATION

Davies and Brooke (1988) focused attention on the evolutionary arms race between bird brood parasites (cuckoos) and their hosts. Examples of their findings include that selection favored egg mimicry by the cuckoos, so that eggs matched their host's eggs, seasonal timing of laying that matched the host's timing, and laying eggs at the right time of day for acceptance. For each evolutionary measure employed by cuckoos, their reed warbler host species had countermeasures (Stoddard and Kilner, 2013).

This game of evolutionary leapfrog, in which measure meets countermeasure is characterized as an evolutionary arms race. Biologists had long been aware of these arms races, and the scientific formulation of arms races came with Ehrlich and Raven's (1964) classic paper on coevolution in butterflies as well as van Valen's Red Queen model (see Chapter 40: The Red Queen). Davies and Brooke (1988) placed the focus on a coevolutionary arms race between parasite and host.

One very interesting point made by Davies and Brooke (1988) was that the obvious solution for reed warblers—expelling the parasite's eggs—carries with it a variety of costs. These include mistakes made in egg identification, as a situation is created in which selection favors egg mimicry. The better the egg mimesis, the more likely a misidentification will be made, and the higher the potential cost for the host becomes.

Conceptual Breakthroughs in Ethology and Animal Behavior.
DOI: http://dx.doi.org/10.1016/B978-0-12-809265-1.00062-9

They also addressed the question of why selection has favored mimic eggs while cuckoo chicks have such different appearances and behaviors than the chicks of their warbler hosts. Ultimately the answer to this question lies in the importance of competition among chicks within a nest, in which larger and faster-growing parasite chicks have an advantage. Animal sensory systems are often tuned to respond more strongly to higher-magnitude stimuli, thus large parasite chicks become a supernormal stimulus that attracts parental attention, rather than causing rejection.

These concepts about how animal's sensory systems can lead them to counterintuitive behavioral choices later came into play with Christy's (1995) sensory trap hypothesis. When one animal uses a signal that is already attractive to another, but in a different context, then it can lure in its target. Christy (1995) was referring to mating attraction and sexual selection, but clearly the behavior of a cuckoo chick can be seen as a sensory trap used to ensnare parental care from the wood warbler host.

IMPACT: 3

Davies and Brooke's (1988) study broke new and important ground by taking a well-known system and subjecting it to rigorous experimental dissection. Through their experiments they established the adaptive value of each strategy in an intricate evolutionary arms race. In addition to stimulating a generation of work on host–parasite coevolution, this paper is a model for how experimental biology should be done, with clearly expressed hypotheses and cleanly designed experiments.

SEE ALSO

Chapter 16, 1947 The Evolution of Clutch Size; Chapter 40, The Red Queen.

REFERENCES AND SUGGESTED READING

Christy, J.H., 1995. Mimicry, mate choice, and the sensory trap hypothesis. Am. Nat. 146, 171–181.

Davies, N.B., de, L., Brooke, M., 1988. Cuckoos versus reed warblers: adaptations and counteradaptations. Anim. Behav. 36, 262–284.

Dawkins, R., Krebs, J.R., 1979. Arms races between and within species. Proc. Royal Soc. London B: Biol. Sci. 205, 489–511.

Ehrlich, P.R., Raven, P.H., 1964. Butterflies and plants: a study in coevolution. Evolution 18, 586–608.

Stoddard, M.C., Kilner, R.M., 2013. The past, present and future of "cuckoos versus reed warblers." Anim. Behav. 85, 693–699.

Chapter 63

1990 Fear

THE CONCEPT

Fear is the subjective representation of potential danger in the environment. Fear can be protective, in that it keeps animals from harm, or damaging, if excessive fear prevents necessary maintenance, mating, or care for young.

THE EXPLANATION

The great Princeton population biologist, Robert MacArthur, influenced the creation of models of population dynamics by helping us to understand niches and interactions among species occupying different niches. His work led to work that helped ecologists to precisely predict the fluctuations of predator and prey populations as well as the influence of parasites and disease on population cycles. But older models assumed that experience did not affect the behavior animals in prey populations. What if the perceived presence of predators, or previous experience of observing predation events, caused prey to become fearful and consequently more elusive?

In their seminal paper, Lima and Dill (1990) considered just that. Animals like deer are fully capable of adjusting their behavior based on the presence of predators, such as mountain lions (Brown et al., 1999). Being less fearful is advantageous when predators are absent, as a herbivore can unmindfully forage under these conditions, but invoking fear when predators are present has great benefits. This brings a nonlinear element to prey responses that, in the view of Brown and his colleagues (1999) helps to explain the rarity of large predators; their mere presence makes their prey seem less abundant.

The acknowledgement of fear in animals fits well with a shifting view of the emotional lives of animals that comes up again in Chapter 74: 2000 Emotion and the Brain, and in Chapter 77: 2004 Behavioral Syndromes—Personality in Animals. While owners of companion animals such as dogs or cats have always known that associations with pain or surprise can generate a fearful response in their animals, it is interesting that behavioral ecologists were so late in recognizing the importance of fear in shaping the behavior of

Conceptual Breakthroughs in Ethology and Animal Behavior.
DOI: http://dx.doi.org/10.1016/B978-0-12-809265-1.00063-0

many wild animals. The reluctance to recognize this mechanism may have stemmed from a desire to not anthropomorphize study animals.

In laboratory rodents, behavioral tests that give animals the option of moving into less secure locations (open-field testing) have allowed the discovery of the pharmacology of fearful behavior as well as the genetics that underlie differences in fearfulness among animals (see Chapter 61: 1985 An Animal Model for Anxiety). One very interesting outcome of open-field tests of rodents is that animals vary in their genetic susceptibility to anxiety, a fear-related phenomenon. Thus fear of a location or a social context may be learned, but variation in how these animals respond to fear-generating experiences is likely genetic.

IMPACT: 5

Consideration of fear in animals was a groundbreaking moment in thinking about animal emotions. It also led to the broader question of how animals vary within species in their responses to risky or stressful situations. While the specific topic of fear has been subsumed into a broader conversation about emotion and personality in animals, the Lima and Dill (1990) paper was a key stimulus that started animal behaviorists thinking about these issues.

SEE ALSO

Chapter 25, 1960 Motivation and Drive; Chapter 27, 1964 Dopamine and Reward Reinforcement; Chapter 51, 1978 Animal Models for Depression; Chapter 61, 1985 An Animal Model for Anxiety; Chapter 65, 1991 Pain in Animals; Chapter 77, 2004 Behavioral Syndromes—Personality in Animals; Chapter 74, 2000 Emotion and the Brain; Chapter 55, 1980 The Risk Paradigm.

REFERENCES AND SUGGESTED READING

Boissy, A., 1995. Fear and fearfulness in animals. Quart. Rev. Biol. 70, 165–191.

Brown, J.S., 2010. Ecology of fear. In: Breed, M.D., Moore, J. (Eds.), The Encyclopedia of Animal Behavior. Elsevier Ltd., Oxford, pp. 581–587.

Brown, J.S., Laundre, J.W., Garung, M., 1999. The ecology of fear: Optimal foraging, game theory, and trophic interactions. J. Mammal. 80, 385–399.

Lima, S.L., Dill, L.M., 1990. Behavioral decisions made under the risk of predation: a review and prospectus. Canadian J. Zool. 68, 619–640.

Olsson, A., Phelps, E.A., 2007. Social learning of fear. Nat. Neurosci. 10, 1095–1102.

Chapter 64

1990 The Challenge Hypothesis

THE CONCEPT

Testosterone levels show large variations over time and among individuals. Testosterone levels are determined by three factors: the physiological baseline, cyclical (often seasonal) reproductive patterns, and responses to social environment or challenges. Understanding the regulation of testosterone in this manner helps to explain its highly variable expression.

THE EXPLANATION

Human knowledge that the testicles affect male behavior and aggression extends back before our written record. Removal of testicles—known as castration, gelding, or caponization—plays a key role in making male domestic animals behaviorally manageable and making their meat palatable. The identification of testosterone as the primary active product of the testes was an early goal of endocrinologists and Adolf Butenandt received the 1939 Nobel Prize in chemistry for working out its structure.

Subsequent to the purification and characterization of testosterone, techniques were developed to measure serum levels of the hormone and ultimately comparisons of testosterone levels became common. These comparisons followed three primary lines—within species through an animal's lifespan, among individuals within a species, and between species. One fascinating outcome of these studies was that testosterone levels are highly variable, even when developmental and seasonal factors are controlled.

This observation led Wingfield and his colleagues to develop the challenge hypothesis (Wingfield et al., 1990). Working from the knowledge that a basal level of testosterone is necessary for metabolic function, and that in most adult male birds testosterone increases during the breeding season, Wingfield et al. (1990) developed a model that explains that the remaining variation in testosterone levels results from social challenges to male status. In particular polygynous birds, which tend to be in constant conflict with other males and which are unlikely to be involved in parental care, have high levels of testosterone (see Chapter 38: 1973 Game Theory).

Conceptual Breakthroughs in Ethology and Animal Behavior.
DOI: http://dx.doi.org/10.1016/B978-0-12-809265-1.00064-2

Their testosterone levels are also responsive to their social challenges. Monogamous male birds, on the other hand, have relatively low testosterone levels and their levels do not rise so readily in response to social challenges. Because high testosterone levels impose high metabolic costs and aggression carries risk of injury, the challenge hypothesis explains much about the trade-offs involved in monogamous versus polygynous mating systems.

Wingfield's work also helped to place emphasis on how an animal's external social environment and their internal hormonal environment interact. At least indirectly, the challenge hypothesis caused field biologists to think about how levels of hormones other than testosterone vary over time. This scientific mindset has come strongly into play in studies of corticosteroid—stress hormone—responses to social and environmental challenges faced by animals.

IMPACT: 2

Wingfield et al.'s (1990) paper was the first to integrate how testosterone both regulates behavior and is regulated by behavior. While they are clear in stating that this hypothesis stems from studies of birds, the theory is broadly applicable in vertebrates, in which testosterone regulates male sexual behavior and aggression, and has had considerable influence on field studies of behavioral endocrinology as well as on our understanding of how captivity influences testosterone levels.

SEE ALSO

Chapter 18, 1948 Hormones and Behavior; Chapter 50, 1977 The Evolution of Mating Systems.

REFERENCES AND SUGGESTED READING

Bell, A.M., 2007. Future directions in behavioural syndromes research. Proc. Royal Soc. B-Biol. Sci. 274, 755–761.

Goymann, W., Landys, M.M., Wingfield, J.C., 2007. Distinguishing seasonal androgen responses from male-male androgen responsiveness: revisiting the challenge hypothesis. Horm. Behav. 51, 463–476.

Ketterson, E.D., Nolan, V., 1999. Adaptation, exaptation, and constraint: a hormonal perspective. Am. Nat. 154, S4–S25.

Ketterson, E.D., Nolan, V., Wolf, L., Ziegenfus, C., 1992. Testosterone and avian life histories: effects of experimentally elevated testosterone on behavior and correlates of fitness in the dark-eyed junco (*Junco hyemalis*). Am. Nat. 140, 980–999.

Wingfield, J.C., Hegner, R.E., Dufty, A.M., Ball, G.F., 1990. The "challenge hypothesis": theoretical implications for patterns of testosterone secretion, mating systems, and breeding strategies. Am. Nat. 136, 829–846.

Wingfield, J.C., Lynn, S.E., Soma, K.K., 2001. Avoiding the "costs" of testosterone: ecological bases of hormone-behavior interactions. Brain Behav. Evol. 57, 239–251.

Chapter 65

1991 Pain in Animals

THE CONCEPT

Pain is a subjective representation associated with damage to body tissues and inflammation. Nonhuman animals perceive and respond to pain.

THE EXPLANATION

As inconceivable as it might sound from a 21st-century perspective, decades ago many scientists and nonscientists firmly held the belief that pain was a subjective representation of a feeling that belonged uniquely to humans. Even though everyday experience with companion animals, domesticated farm animals, and animals targeted in hunting would tell a reasonable person that dogs, horses, chickens, and deer, just to give a few examples, all feel pain when injured, philosophically this was a difficult point for many people to accept. My own first deep impression of animal pain came from the screams of a rabbit that a playmate had shot with an arrow. From that moment, for me, there has been no question that pain is not a philosophical concept but rather a biological property held in common among many, perhaps all, animals.

Given this background it should not be too surprising that Patrick Bateson's paper (1991) on animal pain was necessary. As a response to the resistance to acknowledging animal pain, this prominently placed publication brought attention to both thinking about pain in animals and about how to measure that pain.

Notable in any consideration of pain in animals is the remarkable stoicism with which some animals cope with pain. Part of the difficulty in assessing and acknowledging pain in nonhuman animals is our inability to have a language-based conversation with the animal about how it is feeling. This makes a visit to a veterinarian more like taking an infant to a pediatrician than an adult human visiting an internist. A skilled veterinarian knows where to touch your dog and how to move its joints in order to assess very real sensations like arthritic pain. A dog may not flinch when it walks, limp, or whimper but that does not mean pain is absent. Animal stoicism may have

Conceptual Breakthroughs in Ethology and Animal Behavior.
DOI: http://dx.doi.org/10.1016/B978-0-12-809265-1.00065-4

evolved as a means of maintaining social rank, as perceived weakness can play out poorly in some animal societies.

Robert Elwood (e.g., Sneddon et al., 2014) has led the way in testing hypotheses of pain in a wide variety of animals, including invertebrates. In addition to insightful behavioral observations that help to identify pain, investigators use analgesics (painkillers) and other pharmacological methods to develop convincing evidence for the generality of pain as a mechanism in animals. The point of view developed in Elwood's (2012) work suggests that the question at this point is not so much whether a lobster feels pain as it is cooked, but instead whether a human's moral compass is willing to accept an animal's pain as part of the price of the meal.

Dissenting voices in the scientific community (e.g., Rose et al., 2014) have argued that the assertions for perception of pain in animals have over-reached the data, and that at least in fish the neurons typical of pain perception (a particular type of nociceptor) are absent and that behavioral expression in fish following surgeries does not reflect the presence of pain. Rose et al. (2014) extend their arguments to suggest that the work on crustacean pain is also flawed. As with any scientific question, there is room for discussion, argument, and examination of experimental methods, but the overall principle, that pain is within the sensory realm of many animals, seems to be firmly established.

IMPACT: 8

The acknowledgment that pain can exist in animals, including possibly those that are far from humans on the evolutionary tree, has extreme importance in shaping our practices in the treatment of companion animals, farm animals, and wildlife. This adds to the societal debate that has raged over the last 50 years about animal rights and human responsibilities to animals. Animal behaviorists have a special role in helping to interpret whether behavior is an expression of pain and to propose approaches to animals that minimize the prospects that they will be subjected to pain. Much remains to be discovered about animal perceptions of pain.

SEE ALSO

Chapter 11, 1894 Morgan's Canon.

REFERENCES AND SUGGESTED READING

Bateson, P., 1991. Assessment of pain in animals. Anim. Behav. 42, 827–839.

Elwood, R.W., 2012. Evidence for pain in decapod crustaceans. Anim. Welfare 21, 23–27.

Rose, J.D., Arlinghaus, R., Cooke, S.J., Diggles, B.K., Sawynok, W., Stevens, E.D., et al., 2014. Can fish really feel pain? Fish Fisheries 15, 97–133.

Sneddon, L.U., 2015. Pain in aquatic animals. J. Exp. Biol. 218, 967–976.

Sneddon, L.U., Elwood, R.W., Adamo, S.A., Leach, M.C., 2014. Defining and assessing animal pain. Anim. Behav. 97, 201–212.

Chapter 66

1991 Receiver Psychology

THE CONCEPT

In studies of the evolution of animal signaling much emphasis was put on the signals, but not enough attention was paid to the receiving animal. That animal's ability to detect signals, discriminate among signals, and remember signals play key roles the evolution of signals.

THE EXPLANATION

Sending and receiving messages is a collaboration between the sender and the receiver. Both have to carry the data, either learned or genetic, to encode and decode the information to be embodied in the signal. The sender must have the ability to create the signal and the receiver must be able to detect the signal. The receiver's capacity to discriminate among potential signals is a key initial step in decoding signals. Finally, the receiver's memory, either short-term or consolidated, allows the use of the information gained from the signal.

Guilford and Dawkins (1991) established these points as key elements of receiver psychology at a time when the focus of investigations on animal signaling was biased to looking at the sender and the evolutionary forces impacting the sender. They brought a balanced focus to communication studies that has served animal behaviorists well.

Another way of looking at signal evolution that stems from Guilford and Dawkins (1991) is to look at the process as coevolutionary, analogous to the coevolutionary models used to explain plant–pollinator relationships or predator–prey relationships. In pollination systems, a plant species may evolve in ways that ensure pollinator fidelity by limiting the number of pollinator species that can access the rewards with flowers. This floral evolution then places evolutionary pressure on pollinator species for features that allow them to work that kind of flower. Similarly, fast predators force evolution of faster prey, which in turn stimulates evolution of even faster predators. See Chapter 40: 1973 The Red Queen, for further explanation of this type of evolutionary model.

Conceptual Breakthroughs in Ethology and Animal Behavior.
DOI: http://dx.doi.org/10.1016/B978-0-12-809265-1.00066-6

Evolutionary pressures on receiver sensory and discriminatory abilities has resulted in remarkable adaptations, including those allowing perception of electrical signals in water by electric fish, ultrasounds by mice, and infrasounds by elephants. These perceptive abilities were not "invented" for the purpose of communication, as evolution almost always builds new and refined abilities from preexisting devices.

Rowe (2013) gives an update on the concepts underlying receiver psychology. Looking back at the impact of the original paper on two decades of research, one very key point was the distinction made by Guilford and Dawkins between detectability—the design of signals so that receivers can pick them out from the multitude of noise in an environment—and efficacy—the efficiency with which the message is delivered. From a receiver's point of view, redundancy can be key to efficacy, lending support to the importance of multimodal signals (see Chapter 73: 1999 Multimodal Communication).

IMPACT: 4

The Guilford and Dawkins (1991) paper caused important changes in how signal evolution is interpreted. The paper is foundational to contemporary studies of animal signaling.

SEE ALSO

Chapter 44, 1974 Parent–Offspring Conflict; Chapter 50, 1977 The Evolution of Mating Systems; Chapter 73, 1999 Multimodal Communication.

REFERENCES AND SUGGESTED READING

Candolin, U., 2003. The use of multiple cues in mate choice. Biol. Rev. 78, 575–595.

Endler, J.A., 1992. Signals, signal conditions, and the direction of evolution. Am. Nat. 139, S125–S153.

Endler, J.A., Basolo, A.L., 1998. Sensory ecology, receiver biases and sexual selection. Trends Ecol. Evol. 13, 415–420.

Guilford, T., Dawkins, M.S., 1991. Receiver psychology and the evolution of animal signals. Anim. Behav. 42, 1–14.

Guilford, T., Dawkins, M.S., 1993. Receiver psychology and the design of animal signals. Trends Neurosci. 16, 430–436.

Rowe, C., 1999. Receiver psychology and the evolution of multicomponent signals. Anim. Behav. 58, 921–931.

Rowe, C., 2013. Receiver psychology: a receiver's perspective. Anim. Behav. 85, 517–523.

Chapter 67

1992 Working Memory

THE CONCEPT

There are different types of Short-Term Memory (STM), of which working memory is one. Working memory facilitates cognitive functions such as understanding language and reasoning. As a type of working memory, STM gives animals a way of storing important but transitory information for a limited amount of time.

THE EXPLANATION

Metaphorically, working memory is a place where intermediate steps in the formation of syntax occurs while a sentence is produced, where the intermediate results in an arithmetic problem sit, and where hypotheses are stored and then accepted or dismissed. Working memory is essential to create a sentence or to solve a problem. Working memory is about the process to get to a desired point and then moving on. Knowledge stored in working memory is highly transitory as retention over the long-term would merely be mental clutter.

Baddeley (1992) describes three sorts of working memory. The first of these is a central executive function controlling attention and maintenance of focus on a problem. Two subsidiary functions, one that processes visual images and another that helps in sentence construction, complete the triad of working memory. This concept is truly metaphorical as no specific brain location or circuitry is associated with either of these memory types. Baddeley (1992) argues that the working memory concept supplants STM, but the descriptions of the functions of these two memory types are similar, with STM serving as a register used in counting tasks or storing a simple sequence.

Baddeley's (1992) work builds on a key question developed by Miller (1956); how many discrete chunks of information can be stored in STM at a given time? Cowan's (2001) review consolidates and builds on the concepts developed by Miller and Baddeley. Miller's (1956) paper is a key moment in

Conceptual Breakthroughs in Ethology and Animal Behavior.
DOI: http://dx.doi.org/10.1016/B978-0-12-809265-1.00067-8

the literature on STM, and Cowan's (2001) review could have equally well been chosen as the moment to highlight in this book.

Most of the consideration of this question has been shaped around studies of humans, who, conveniently, can be easily asked to report on what they remember. The experiences of human subjects can be manipulated in myriad ways, so that suppositions about learning and memory can be tested multiple times from different angles. Testing STM in animals is more challenging, largely because of the difficulty in getting an animal to report what it remembers.

Miller's (1956) central conclusion, that STM can hold more or less seven elements, or chunks, is supported in general terms by much of the literature that followed the publication of his work. Cowan argues that the magical number may be four, rather than seven; this is worth considering but should not overshadow the general conclusion that STM holds a small number of discrete items.

The relevance of this information about STM to the behavior of animals, other than humans, is that when animals are given a counting task, they are often proficient up to four to seven items. Beyond that, the assessment seems to become few versus many, rather than an accurate count. Investigators have taken this to suggest that the counting limitations of animals corresponds with the STM limitation on chunks of information; an animal is more likely to have evolved the ability to use STM to count small numbers than to have evolved cognitive abilities to perform larger arithmetic problems.

Cowan (2001) correctly points out that it is difficult to know exactly what a chunk is, where the chunk is physically located in the memory, and whether differences in assessments of the number of chunks stored arise from context (such as type of information), methodological differences, or when nonhumans are tested, species differences. In fact, some investigators continue to doubt the existence of STM as a separate mechanism of memory.

STM remains a very important area of research, and its relationship to counting and other simple assessments that nonhuman animals might do is a very open question. For example, it would be very interesting to know how a mechanism like STM functions in the process by which coots reject eggs laid by nest parasites (Lyon, 2003; Shizuka and Lyon, 2010). Coots can assess both the number of eggs present and their hatching order, and use this information to reject eggs laid by other birds. Are the coots limited in their ability to count eggs by the number of STM chunks available? Answering this type of question will help to establish the role of STM in the behavior of animals in field settings.

IMPACT: 7

Memory is highly complex and our interpretations of memory processes are highly based on subjective human experiences of our own memory processes. Hypotheses like working memory and STM originate in human efforts to consciously monitor how our minds work and to report the sequences we experience. This approach is bolstered by the observation that neurological conditions, like Alzheimer's disease, specifically impair the expression of the results of working memory and STM. Of course, we do not know whether the memory processes themselves are damaged or whether the problem lies in neural locations for executive functions.

SEE ALSO

Chapter 19, 1948 Information Theory; Chapter 37, 1973 Episodic Memory; Chapter 52, 1978 Theory of Mind; Chapter 54, 1980 Semantic Communication.

REFERENCES AND SUGGESTED READING

Baddeley, A., 1992. Working memory. Science 255, 556–559.

Cowan, N., 2001. The magical number 4 in short-term memory: a reconsideration of mental storage capacity. Behav. Brain Sci. 24, 87–114.

Lyon, B.E., 2003. Egg recognition and counting reduce costs of avian conspecific brood parasitism. Nature 422, 495–499.

Miller, G.A., 1956. The magical number seven, plus or minus two: some limits on our capacity for processing information. Psychol. Rev. 63, 81–97.

Shizuka, D., Lyon, B.E., 2010. Coots use hatch order to learn to recognize and reject conspecific brood parasitic chicks. Nature 463, 223–226.

Chapter 68

1994 Ecosystem Engineers

THE CONCEPT

Physical modifications of the environment by animals, such as beaver building a dam, affect ecosystem properties like nutrient cycles, the distribution of soil particles and chambers in soil, and water availability.

THE EXPLANATION

How do the engineering efforts of animals impact the abiotic world around them? Animals move soil, build structures, modify watercourses with dams and other inventions, and selectively remove nutrients from some locations and add nutrients to other spots. In their highly original paper on the subject of ecosystem engineers, Jones and colleagues (1994) said that: "Conspicuously lacking from the list of key processes in most text books is the role that many organism play in the creation, modification and maintenance of habitats." And, indeed, the important effects that organisms have on their habitats had remained unappreciated until the publication of this paper.

From the perspective of animal behavior, ecosystem engineering is a core activity for many species. Many types of animals modify their environment to make it more habitable, to make their foraging more efficient, to attract mates, and to protect themselves from predators.

Beaver suggest to us our most immediate image of ecosystem engineers; their magnificent impoundments in streams and small rivers, create a still water habitat for the construction of their lodge. Other animals, such as prairie dogs, ants, and crawfish, excavate underground tunnel systems for shelter. These excavations unearth soil from subsurface layers and bring it to the surface, provide routes for air and water to reach underground, and oftentimes food or feces deposited in the tunnel system rearranges nutrients in the ecosystem. Changes in waterflow and soil distribution are the two major ways in which animals engineer their ecosystems (Fig. 68.1).

There are also other, more subtle routes by which animals can have major ecosystems effects. For example, ants and termites typically deposit their refuse and feces within a meter or two of their nest. These discards are rich

Conceptual Breakthroughs in Ethology and Animal Behavior.
DOI: http://dx.doi.org/10.1016/B978-0-12-809265-1.00068-X

FIGURE 68.1 A lodge of the American beaver, *Castor canadensis*. Beaver cut stems and branches from live trees, strip the bark, and build elaborate and long-lasting lodges, which protect them from predators and weather. They also construct, using the same process, dams across creeks and streams; these impound the water and create a moat around their lodge. *Photo courtesy Michael Breed.*

in nitrogen and phosphorus, yielding an uneven pattern of nutrients (high around ant mounds, low away from ant mounds) around mound sites. When an ant or termite colony dies, the release of nutrients from the disintegrating mound effectively fertilizes the surrounding area.

The roles that animals play as ecosystems engineers create much of the small-scale variation within ecosystems. Ecosystems engineers, by concentrating water and nutrients and by shifting soil horizons create potential hot-spots for plant and microbial growth. Environmental heterogeneity created by ecosystems engineers supports added layers of biological diversity in many habitats.

IMPACT 6

The overview by Jones et al. (1994) about ecosystems engineering is a great example of the confluence of behavior and ecology, and of how behavior can have unexpected impacts on ecological processes. Animal behaviorists have sometimes endured long waits for ecologists to recognize the importance of behavior in ecological processes, and this publication promoted the overall consideration of behavior as a critical piece of the ecological puzzle.

REFERENCES AND SUGGESTED READING

Crooks, J.A., 2002. Characterizing ecosystem-level consequences of biological invasions: the role of ecosystem engineers. Oikos 97, 153–166.

Folgarait, P.J., 1998. Ant biodiversity and its relationship to ecosystem functioning: a review. Biodiver. Conservat. 7, 1221–1244.

Gutierrez, J.L., Jones, C.G., Strayer, D.L., Iribarne, O.O., 2003. Mollusks as ecosystem engineers: the role of shell production in aquatic habitats. Oikos 101, 79–90.

Jones, C.G., Lawton, J.H., Shachak, M., 1994. Organisms as ecosystem engineers. Oikos 69, 373–386.

Jones, C.G., Lawton, J.H., Shachak, M., 1997. Positive and negative effects of organisms as physical ecosystem engineers. Ecology 78, 1946–1957.

Lavelle, P., Bignell, D., Lepage, M., Wolters, V., Roger, P., Ineson, P., et al., 1997. Soil function in a changing world: the role of invertebrate ecosystem engineers. European J. Soil Biol. 33, 159–193.

Chapter 69

1996 Conservation Behavior

THE CONCEPT

Conservation behavior is the use of results from studies of animal behavior in making decisions about conservation strategies.

THE EXPLANATION

To function well, conservation programs need a thorough knowledge of the behavior of the animals within the ecosystems targeted for conservation. Everything about behavior has conservation implications: Noise levels interfere with the mating systems of birds and frogs. Pollution can change food availability, sometimes in ways that are not immediately apparent. Animals need corridors for dispersal and migration, and long-distance migrants need stopover locations with adequate habitat for protection and feeding between migratory legs (Fig. 69.1).

Nevertheless, behavior is often the poor step-child in conversations about conservation. Managers have tended to focus on land acquisition, recreational opportunities for humans, and if animal behavior was involved, it was often on the level of keeping tourists and animals apart. Conservation scientists often delve into landscape level processes, climate change, and macronutrient flows (such as nitrogen and phosphorus) without narrowing their focus to the conservation problems of individual species. Conservation behavior brings a needed voice to the table in conservation management decisions.

Eberhard Curio (1996) was not the first ethologist or behavioral ecologist to notice that animal behavior often figured so little in conservation plans, but his publication sparked an ongoing debate among the community of scientists about the role of behavior in conservation planning. Some progress has been made since Curio (1996), in terms of reserve designs that incorporate plans for meeting the behavioral needs of animals, but much remains to be done in this field.

Practical methods for assessing how much human movements disturb animals have been developed. For critical species, the intricacies of achieving matings in captivity, rearing offspring in captivity, and releasing animals

Conceptual Breakthroughs in Ethology and Animal Behavior.
DOI: http://dx.doi.org/10.1016/B978-0-12-809265-1.00069-1

FIGURE 69.1 John James Audubon's portrait of the now-extinct passenger pigeon. This bird was once one of the most common birds in the eastern United States. Knowledge of animal behavior can help to prevent future such extinctions. *Plate 62 of Birds of America by John James Audubon depicting Passenger Pigeon,1827–1838, John James Audubon (1785-1851), University of Pittsburgh.*

into the wild have been mastered. Techniques for managing how wildlife and humans interact in our ever-growing cities and suburbs are being improved. A textbook by Blumstein and Fernandez-Juricic (2011) captures the current status of the field in a concise and engaging manner.

One bump in the road was a critique by Caro (2007), which questioned whether behavior could or should be integrated in conservation plans. This stimulated strong responses, including one from Buchholz (2007). The Caro (2007) paper generated a sharpening of focus and a stronger sense of purpose among behaviorists who are dedicated to conservation.

IMPACT: 9

Conservation behavior is of prime importance in making wise conservation investments. Behavior, including topics such as foraging, food choice, mating systems, dispersal, and migration, can inform many conservation decisions. In an era when science is looking to justify itself, applying knowledge from studies of animal behavior to conservation issues is essential to the survival of the field.

SEE ALSO

Chapter 6, 1800s Birds in Their Natural Setting; Chapter 7, 1800s The Great Explorers.

REFERENCES AND SUGGESTED READING

Berger-Tal, O., Polak, T., Oron, A., Lubin, L., Kotler, B.P., Saltz, D., 2011. Integrating animal behavior and conservation biology: a conceptual framework. Behav. Ecol. 22, 236–239.

Blumstein, D.T., Fernandez-Juricic, E., 2010. A Primer of Conservation Behavior. Sinauer Associates, Sunderland, MA.

Buchholz, R., 2007. Behavioral biology: an effective and relevant conservation tool. Trends Ecol. Evol. 22, 401–407.

Caro, T., 2007. Behavior and conservation: a bridge too far? Trends Ecol. Evol. 22, 394–400.

Curio, E., 1996. Conservation needs ethology. Trends Ecol. Evol. 11, 260–263.

Chapter 70

1996 The Molecular Basis of Learning

THE CONCEPT

A molecular mechanism for long-term storage learned information involves protein synthesis and structural changes in the neurons required to maintain the information. This is true of both explicit memory—information about events and locations—and implicit memory—learned procedures and associations. This mechanism is evolutionarily conserved and remarkably similar in form in all animals.

THE EXPLANATION

Eric Kandel's pursuit of the physiological and molecular bases of memory began in the 1950s. He single-mindedly pursued the fundamental mechanisms of memory and in the course of his work developed unique model systems and key techniques, first in neurophysiology and later in molecular neurogenetics. He is famed for exploiting the very simple nervous system and behavioral responses of sea slugs, *Aplysia*, in sophisticated studies of learning and memory.

It is difficult to pin down a specific finding from Kandel and his research group that towers in importance over other findings. Each publication builds from previous work and it is the body of evidence that convinces, rather than any specific insight that rises above the others. Kandel himself (e.g., Bailey et al., 1996) points to the development of knowledge about behavioral aspects of the *Aplysia* system (Frost et al., 1985) as the key moment in his work.

However, I have chosen 1996 as the pivot point for the influence of his work, as this was the year in which the work on the molecular genetics of learning seemed to come together in an integrated picture with two publications, Bailey et al. (1996) and Mayford et al. (1996). These establish the importance of the CaMKII gene in the sequence of events leading to memory formation. This gene, which is a protein kinase that depends on the presence

Conceptual Breakthroughs in Ethology and Animal Behavior.
DOI: http://dx.doi.org/10.1016/B978-0-12-809265-1.00070-8

of calcium (via the gene calmodulin) for its activity, specifically facilitates the protein synthesis necessary for learning. CaMKII is one of the most abundant proteins in the brain.

Kandel shared the 2000 Nobel Prize for Physiology or Medicine for work on signaling in the nervous system. His work stands above the field in its clarity, creativity, and contribution to understanding a fundamental mechanism of life.

IMPACT: 8

Kandel's findings were certainly expected. Of course learning and memory has a molecular basis an enormous level of care and insight was required to work out the details. In study after study, Kandel provided ingenious insights into one of the basic mechanisms of animal life. The impact of his work is not so much in establishing that a phenomenon exists as in serving as a model of the very best kind of careful science.

SEE ALSO

Chapter 13, 1938 Skinner and Learning; Chapter 34, 1971 Behavioral Genetics; Chapter 42, 1974 *Caenorhabditis elegans* Behavioral Genetics; Chapter 68, 1992 Working Memory; Chapter 75, 2000 Social Amoebas and Their Genomes.

REFERENCES AND SUGGESTED READING

Bailey, C.H., Bartsch, D., Kandel, E.R., 1996. Toward a molecular definition of long-term memory storage. Proc. Natl. Acad. Sci. USA 93, 13445–13452.

Frost, W.N., Castellucci, V.F., Hawkins, R.D., Kandel, E.R., 1985. Monosynaptic connections made by the sensory neurons of the gill-and siphon-withdrawal reflex in *Aplysia* participate in the storage of long-term memory for sensitization. Proc. Natl. Acad. Sci. USA 82, 8266–8269.

Goelet, P., Castellucci, V.F., Schacher, S., Kandel, E.R., 1986. The long and the short of long-term-memory: a molecular framework. Nature 322, 419–422.

Kandel, E.R., 2001. Neuroscience-the molecular biology of memory storage: a dialogue between genes and synapses. Science 294, 1030–1038.

Kandel, E.R., Spencer, W.A., 1968. Cellular neurophysiological approaches in study of learning. Physiol. Rev. 48, 65–134.

Lisman, J.H., Schulman, H., Cline, H., 2002. The molecular basis of CaMKII function in synaptic and behavioural memory. Nat. Rev. Neurosci. 3, 175–190.

Mayford, M., Bach, M.E., Huang, Y.Y., Wang, L., Hawkins, R.D., Kandel, E.R., 1996. Control of memory formation through regulated expression of a CaMKII transgene. Science 274, 1678–1683.

Chapter 71

1998 Self-Organization of Social Systems

THE CONCEPT

Self-organizing social systems operate through complex social mechanisms that arise from simple rules contained by each animal in the group. There is no need for an organizer or for instructions.

THE EXPLANATION

Animal societies sometimes resemble the workings of an elaborately crafted mechanical clock, in which each gear meshes perfectly with its mate, the balance wheel maintains a perfect rhythm, and a spring gives the energy needed to keep the works going. All is efficient, organized, and well designed to communicate the passage of time. As with the cogs, springs, and wheels in a clock or watch, each animal in the society can seem to perform its role perfectly.

But do animal societies really function like clockwork? To consider this question Eric Bonabeau (1998) wondered whether animal societies always need top-down direction in order to function. To continue the metaphor of the clock, does an animal society need an animal to function like the clockmaker, directing the alignment of the gears, adjusting the balance wheel and winding the spring? Calling the reproductive female in an ant colony the "queen" invites exactly this view of the society—that a central organizer is required.

What if each animal in the society carries with it a few simple rules about how to behave? The rules could include how to behave when the colony is short on food, how to behave when the colony is attacked, and so on. The application of these rules, by individual animals, could lead to an organized society without the need for centralized direction.

To extend the metaphor just one step further, a self-organizing society could be visualized as a cartoon, in which the gears, balance wheel, and

Conceptual Breakthroughs in Ethology and Animal Behavior.
DOI: http://dx.doi.org/10.1016/B978-0-12-809265-1.00071-X

other clock parts have sprung legs and are running around finding the appropriate position for their work in the larger mechanism of the clock.

This model of self-organization becomes more efficient if different animals in the society have different thresholds for responding to tasks. Suppose animal A is most responsive to young that need feeding, while animal B is most responsive to construction tasks. Either animal can do both jobs, but animal A is most likely to feed young, while B joins in that task only when unmet need for feeding young creates an emergency. In this way all tasks are covered, but specialists do most of the work on each task. Specialization adds efficiency if the specialists are better, either through learning or physical capability, at doing their task than nonspecialists.

Bonabeau's thinking was embraced by researchers of social insect behavior. Wasps, bees, ants, and termites have highly efficient societies in which there is no clear direction or supervision that organizes the labor. "Queen" and "King" are misnomers in the social insect world, as these animals are merely reproductive specialists, not CEO's for the colony corporation.

R. E. Page employed the response threshold model to delve into the organization of foraging behavior in honeybees (e.g., Page and Erber, 2002) and J. H. Fewell (Jeanson et al., 2007) led the way in using small groups of insects, such as founding queens in ant societies, as a window into understanding self-organization.

Viewing animal societies—not just social insects but also societies as diverse as acorn woodpeckers, wolves, and primates—as at least partial outcomes of self-organization has been fruitful. This view creates a way of analyzing the evolutionary balance between the possible benefits of top-down control and central organization versus the benefits of efficiency and immediacy of response gained from self-organization.

IMPACT: 6

Self-organization is a seminal idea in research on social evolution. It gives insight into the functional aspects of social evolution in the way that kin selection shows us the way in finding selective forces underlying social evolution (see Chapter 28: 1964 Inclusive Fitness and the Evolution of Altruism).

SEE ALSO

Chapter 3, 1623 Social Behavior; Chapter 9, 1859 Darwin and Social Insects; Chapter 28, 1964 Inclusive Fitness and the Evolution of Altruism; Chapter 35, 1971 Reciprocal Altruism; Chapter 36, 1971 Selfish Herds; Chapter 45, 1975 Group Selection.

REFERENCES AND SUGGESTED READING

Beshers, S.N., Fewell, J.H., 2001. Models of division of labor in social insects. Ann. Rev. Entomol. 46, 413–440.

Bonabeau, E., 1998. Social insect colonies as complex adaptive systems. Ecosystems 1, 437–443.

Jeanson, R., Fewell, J.H., Gorelick, R., Bertram, S., 2007. Emergence of increased division of labor as a function of group size. Behav. Ecol. Sociobiol. 62, 289–298.

Page, R.E., Erber, J., 2002. Levels of behavioral organization and the evolution of division of labor. Naturwissenschaften 89, 91–106.

Chapter 72

1998 Gaze Following

THE CONCEPT

Gaze following is watching another animal's eyes in order to determine what it is looking at. Gaze following may require a degree of cognition, as it could reflect the ability to distinguish self from other.

THE EXPLANATION

"What are you looking at?" This simple, often-used, question in human conversation can express curiosity, paranoia, or fear. Carried within the query is our certain knowledge that we can tell how intently someone else is staring at something, and that we can follow their gaze. In other words, we extrapolate from where they stand or sit to where they are looking by following the angle of their eyes. A gaze at a bird in a tree makes us curious—where exactly is the bird? A gaze at a noise in the bushes may make us fearful—is it a tiger? A soft gaze directed at you is attentive, but a hard gaze at you likely feels aggressive and makes you anxious or paranoid. There is so much information to be gleaned from where other animals in a social group look, it would seem surprising if this capability is not shared among all social species.

Yet, until fairly recently, only humans were credited with the ability to follow another animal's gaze. In Michael Tomasello and his colleagues expanded our view of gaze following in a series of experiments that showed that some close relatives to humans—other primate species—could follow gazes of their own species (Tomasello et al., 2008). Couched in the exploration of animals close to humans was the implicit assumption that gaze following requires special cognitive insight that only a human-like animal could have. In the two decades since the primate data became public we have come to understand that gaze following is common among many types of animals, including birds, fish, and reptiles in addition to mammals (Fig. 72.1).

The lateness of acceptance that animals can follow and interpret gazes is part of the larger pattern in animal behavior of attempting to define humans by special, unique characteristics that elevate us above our nonhuman kindred.

Conceptual Breakthroughs in Ethology and Animal Behavior.
DOI: http://dx.doi.org/10.1016/B978-0-12-809265-1.00072-1

FIGURE 72.1 Humans, some primates, dogs, and perhaps other species can gain information by assessing where another animal is looking. Gaze following can be used to access public information, such as the location of a potential predator, or private information, such as a territorial opponent's next action. In human–dog interactions we often point in order to get the dog to attend to a specific object or task. The dog's interpretation of the point is related to gaze following, but requires the dog to recognize that the point is leading to useful information. Some, but not all, dogs can follow a point.

The assumed separateness of humans stems in part from religious traditions that hold humans above all else in the world or view humans as exempt from the baser instincts of animals. Much of the last century of scientific discovery in animal behavior has been about blurring the human/nonhuman line, as so it is with discoveries about gaze following.

One of the more interesting set of studies subsequent to Tomasello et al. (1998) deals with gaze following in wolves and dogs. No doubt that wolves and dogs can follow the gaze of other members of their pack. But are they capable of following the gaze of a human? Some investigators argue that ability to follow a human's gaze is an evolutionary innovation in dogs, associated with their domestication. Others argue that wolves and dogs are equally well equipped to follow human gazes, but that dogs are socialized to take more advantage of information from their codependent species. This extends to the ability to follow a pointing motion by a human, which some dogs but few wolves can do; is following a point an ability unique to dogs or does it depend on socialization? The answers to these questions will help to sort out how dogs and humans became so evolutionarily entwined.

Gaze following remains one of the central issues of animal cognition. How much awareness of self versus other is needed to use the information inherent in the direction another animal is looking? Cognitive science suggests that the evolutionary step to being able to identify self and to assign intentions to others is a high bar that not all animals have evolved to cross (Davidson et al., 2014). Alternatively, is an elaborate cognitive construct essential for gaze following or is it a fairly simple exploitative behavior? Findings of evidence of gaze following in representatives of all the major vertebrate types may suggest the need to downplay cognition as a precondition for gaze following. In reality, the discovery of gaze following and its importance in information gathering is a major leap in the study of animal behavior, quite independent of the question of cognition.

IMPACT: 4

Gaze following is one of the most interesting and controversial pieces of the overall puzzle of animal cognition. The question, e.g., of whether dogs follow human gazes as a consequence of domestication has generated considerable heat in the scientific conversation. Based on the intensity of the controversies and the difficulties involved in the resolution of how and why animals follow each other's gazes, this is a high-impact area of inquiry.

SEE ALSO

Chapter 66, 1991 Receiver Psychology; Chapter 79, 2004 Public and Private Information.

REFERENCES AND SUGGESTED READING

Davidson, S., Butler, E., Fernández-Juricic, E., Thornton, A., Clayton, N.S., 2014. Gaze sensitivity: function and mechanisms from sensory and cognitive perspectives. Anim. Behav. 87, 3–15.

Tomasello, M., Call, J., Hare, B., 1998. Five primate species follow the visual gaze of conspecifics. Anim. Behav. 55, 1063–1069.

Tomasello, M., Carpenter, M., Call, J., Behne, T., Moll, H., 2005. Understanding and sharing intentions: the origins of cultural cognition. Behav. Brain Sci. 28, 675–691.

Tomasello, M., et al. 2008. Origins of human communication. Jean-Nicod lectures. Cambridge, MA: MIT Press.

Chapter 73

1999 Multimodal Communication

THE CONCEPT

The use of more than one sensory mode for a signal, such as communicating the same message simultaneously by both sight and sound, reinforces the message and provides redundancy in case communication in one of the modes is compromised.

THE EXPLANATION

Multimodal signaling is another of the topics discussed in this book which seems obvious to a thoughtful layperson but which took a long time for scientists to embrace (Partan and Marler, 1999). Consider how we recognize another person as an individual. Their facial appearance, voice, odor, and gait probably each works perfectly well. It is handy to have signals of identity in multiple modes; e.g., if it is dark, facial appearance is not available as a cue, but then voice serves equally well. If we try to suppress our odor by showering and using deodorant, then visual and auditory cues are still present. By using multiple modes—methods of communicating the same information—we make sure that our message gets through.

Multimodal signaling involves using more than one of the modalities that animals employ in a coordinated set of signals that conveys information to other animals. The modalities may include visual, auditory, tactile, and chemical cues with which humans are familiar, but could also include magnetic, electrical, or other types of cue that humans cannot perceive or perhaps that we cannot yet imagine. By building a multimodal signal an animal increases the accuracy its communication.

In addition to providing information through alternative cues if one signaling channel is blocked, multimodal signaling provides the advantage of increasing signaling accuracy through redundancy. A quick example will illustrate this. If a visual signal has a 1 in 1000 (0.1%) error rate, and an auditory signal has a similar error rate, the two together are wrong only one

Conceptual Breakthroughs in Ethology and Animal Behavior.
DOI: http://dx.doi.org/10.1016/B978-0-12-809265-1.00073-3

in one-million times. If a third source of information is added, such as odor, with a equal error rate, then the three fail simultaneously only one in one-billion times.

Multimodal signaling also may allow animals to pack more information into a single signaling bout. If the signals are not completely redundant, then each mode may carry additional information. For example, a multimodal alarm signal may carry information about being alert, about the direction of approach of the threat, and perhaps the identity of the threat.

IMPACT: 4

Many animal behaviorists who were working on signaling and communication in the 1990s, when the Partan and Marler paper was published, knew at least intuitively about the importance of multimodal signaling. In fact, composite signals like dogs visually signaling their scent-laden urination locations by raising their leg, had been well studied. Thus bringing attention to the phenomenon had value, but did not impel a new line of investigation. Partan and Marler's (1999) paper was a perspectives note rather than a full-blown paper, but it brings together the ideas that were fermenting at that time about multimodal signaling in a nice, concise way, and consequently it has been well cited.

SEE ALSO

Chapter 66, 1991 Receiver Psychology; Chapter 79, 2004 Public and Private Information.

REFERENCES AND SUGGESTED READING

Bro-Jorgensen, J., 2010. Dynamics of multiple signalling systems: animal communication in a world in flux. Trends Ecol. Evol. 25, 292–300.

Candolin, U., 2003. The use of multiple cues in mate choice. Biol. Rev. 78, 575–595.

Hebets, E.A., Papaj, D.R., 2005. Complex signal function: developing a framework of testable hypotheses. Behav. Ecol. Sociobiol. 57, 197–214.

Kulahci, I.G., Dornhaus, A., Papaj, D.R., 2008. Multimodal signals enhance decision making in foraging bumblebees. Proc. Royal Soc. B: Biol. Sci. 275, 797–802.

Partan, S., Marler, P., 1999. Communication goes multimodal. Science 283, 1272–1273.

Partan, S.R., Marler, P., 2005. Issues in the classification of multimodal communication signals. Am. Nat. 166, 231–245.

Roberts, J.A., Taylor, P.W., Uetz, G.W., 2007. Consequences of complex signaling: predator detection of multimodal cues. Behav. Ecol. 18, 236–240.

Chapter 74

2000 Emotion and the Brain

THE CONCEPT

Emotion is not a function of the primitive brain, sometimes called the limbic system, and can be studied as discrete functions, such as fear, that are assignable to specific anatomical regions of the brain.

THE EXPLANATION

LeDoux's 2000 paper on emotion and the brain is a turning point in understanding animal behavior because he clearly expresses a modern view of the neuroscience of emotions. Many concepts of brain function were developed essentially as metaphors (see Chapter 67: 1992 Working Memory) for poorly understood neural systems. This included emotions, which had been considered by scientists at least as far back as Darwin. Emotions are difficult to define. If they are only considered as subjective experiences they remain the realm of theory and metaphor rather than objective investigation.

LeDoux (2000) focused primarily on fear and the role of the amygdala in regulating fearful responses and organizing memories associated with fear or with trauma that induces fear. LeDoux (2000) also provides an excellent argument for abandoning the concept of the limbic system. The limbic system is the aggregate of the evolutionarily older parts of the brain including the hypothalamus, amygdala, hippocampus, and the olfactory bulbs. Because mammals have the evolutionarily more recent neocortex laid over the limbic system, some influential psychologists believed that primitive noncognitive behavior—impulse and emotion—was based in the limbic system and that mammals had exclusive possession of cognitive abilities residing in the neocortex.

This point of attention relates in interesting ways to Lima and Dill's (1990, see Chapter 63: 1990 Fear) exploration of the evolutionary ecology of fear a decade earlier. It also plays forward to the more recent concept of behavioral syndromes, for which behavior along a shy–bold continuum has been a central research focus (see Chapter 77: 2004 Behavioral Syndromes—Personality in Animals). Worth considering is the idea that fear is somewhat unique

Conceptual Breakthroughs in Ethology and Animal Behavior.
DOI: http://dx.doi.org/10.1016/B978-0-12-809265-1.00074-5

among emotions as its evolutionary value is clear, its ecological consequences are well understood, and we have good knowledge of its neural underpinnings.

Can other emotions, for which characterizations are still more subjective, be resolved in the same way? Love, for example, is an extremely subjective concept and it is hard to know whether the bond a companion animal such as a dog or cat feels with its human is love or exploitation. Nevertheless, social bonding—mate with mate, parent with offspring, offspring with sibs, and so on—is common in the animal world and in vertebrates the roles of vasopressin, oxytocin, and dopamine in bonding are well established. So perhaps love will always remain subjective but bonding can be understood in concrete terms. And it might be so for any emotion that can be semantically disentangled from subjectivity.

Even contemporary scientists sometimes refer to 'higher functions' in the neocortex, despite the fact that we now know that a very broad range of animals have abilities for cognition, calculation, and manipulation even though some lack a neocortex. Current knowledge supports the conclusion that at least some aspects of mental higher functions, including emotion, are present in diverse animal taxa outside the birds and mammals.

IMPACT: 8

LeDoux's (2000) paper is widely cited because it combines well-argued theory about emotions, the limbic system, and cognition with detailed experimental findings concerning fear. It represents a turning point at which a neurobiological approach to a problem in animal behavior came into congruence with thought about the evolution and ecology of behavior.

SEE ALSO

Chapter 27, 1964 Dopamine and Reward Reinforcement; Chapter 37, 1973 Episodic Memory; Chapter 52, 1978 Theory of Mind; Chapter 63; 1990 Fear; Chapter 51, 1978 Animal Models for Depression; Chapter 61, 1985 An Animal Model for Anxiety; Chapter 65, 1991 Pain in Animals.

REFERENCES AND SUGGESTED READING

Brown, J.S., Laundre, J.W., Garung, M., 1999. The ecology of fear: optimal foraging, game theory, and trophic interactions. J. Mammal. 80, 385–399.

Cardinal, R.N., Parkinson, J.A., Hall, J., Everitt, B.J., 2002. Emotion and motivation: the role of the amygdala, ventral striatum, and prefrontal cortex. Neurosci. Biobehav. Rev. 26, 321–352.

LeDoux, J.E., 2000. Emotion circuits in the brain. Ann. Rev. Neurosci. 23, 155–184.

Lima, S.L., Dill, L.M., 1990. Behavioral decisions made under the risk of predation: a review and prospectus. Canadian J. Zool. 68, 619–640.

Phelps, E.A., LeDoux, J.E., 2005. Contributions of the amygdala to emotion processing: from animal models to human behavior. Neuron 48, 175–187.

Young, L.J., Zuoxin Wang, Z., 2004. The neurobiology of pair bonding. Nat. Neurosci. 7, 1048–1054.

Chapter 75

2000 Social Amoebas and Their Genomes

THE CONCEPT

Social amoebas cooperate, compete, and cheat. Genomics gives us a window into the evolution and regulation of these behaviors.

THE EXPLANATION

Slime molds, *Dictyostelium discoideum*, which are really social amoebas, came to the forefront of research on social behavior with the publication of Strassmann et al.'s (2000) analysis of amoeba social behavior. They showed that when a group of amoebas come together to form what is called a slug, some of the amoebas serve to be parts of the nonreproductive stalk, while others migrate to reproductive positions. In other words, there are altruistic amoebas that sacrifice themselves so that other amoebas can reproduce. In further analyses, they found that some amoebas can be "cheaters," exploiting the altruism of others (Fig. 75.1).

Following shortly after Strassmann et al.'s (2000) paper on slime mold social behavior came the thorough analysis of the genome of the social amoeba (Eichinger et al., 2005). Genomics has long held the potential for revolutionizing the study of animal behavior, and genome sequences are available for a wide variety of organisms. The social amoeba stands as a prime example of the potential for genomics can inform our understanding of social behavior.

Honeybees hold similar potential (Weinstock et al., 2006), and as genomes for more animals become available, comparative genomics will yield insights into how behaviors evolved and how they are organized. However, relating results from genomic studies to behavior has proven more complicated than optimistic behavioral geneticists might have predicted in the 1990s, and far more difficult than the genome scientists who did the sequencing in the 2000s would have predicted. Approaches, including those focusing on gene expression, RNA regulation, and proteomics, have given

Conceptual Breakthroughs in Ethology and Animal Behavior.
DOI: http://dx.doi.org/10.1016/B978-0-12-809265-1.00075-7

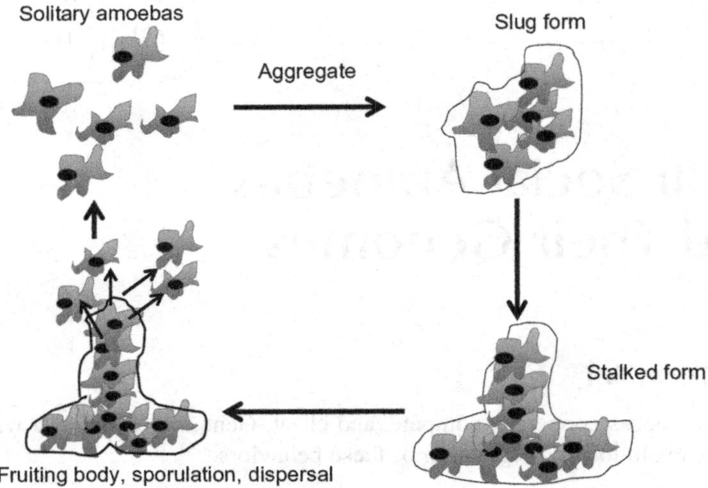

FIGURE 75.1 The social amoeba, *Dictyostelium discoideum*, has a complex life cycle including solitary, sexual, and social phases. This diagram highlights the social phase, in which solitary amoebas aggregate, form a slug, then a stalk, and finally a fruiting body.

interesting insights, but genes turn out to operate in a very tangled web of expression and regulation that makes it nonsensical to search for a single key gene for any one behavior or activity.

Although a single gene for social behavior or a component of social behavior will never be discovered, genes that are essential for social behavior will be revealed by techniques like genetic knockdown, in which the expression of a gene is impaired and its effect can then be assessed. So while a novel set of genes for social behavior does not exist in any organism, major regulatory shifts in social animals will be discovered and the correlation between these shifts and behavior will give insights into how the architecture of gene expression and regulation changes to support social behavior. Social amoebas provide a fascinating model system for this integration of social behavior and genetics. Because they are socially more complex than *Caenorhabditis elegans* (see Chapter 34: 1974 *Caenorhabditis elegans* Behavioral Genetics) they are a better model for social evolution, while *C. elegans* provides a better model for relating the nervous system to behavior.

IMPACT: 7

Social amoebas provide one of the two most elegant and best-developed systems for integrating the study of behavior and genetics at a molecular level; the other such system is *C. elegans* (see Chapter 34: 1974 *Caenorhabditis elegans* Behavioral Genetics). The work on social amoebas sets a trail that research on other organisms should follow.

SEE ALSO

Chapter 28, 1964 Inclusive Fitness and the Evolution of Altruism; Chapter 34, 1971 Behavioral Genetics; Chapter 42, 1974 *Caenorhabditis elegans* Behavioral Genetics.

REFERENCES AND SUGGESTED READING

Eichinger, L., Pachebat, J.A., Glockner, G., et al., 2005. The genome of the social amoeba *Dictyostelium discoideum.* Nature 435, 43–57.

Robinson, G.E., Grozinger, C.M., Whitfield, C.W., 2005. Sociogenomics: social life in molecular terms. Nat. Rev. Genet. 6, 257–270.

Sobotka, J.A., Daley, M., Chandrasekaran, S., Rubin, B.D., Thompson, G.J., 2016. Structure and function of gene regulatory networks associated with worker sterility in honeybees. Ecol. Evol. 6, 1692–1701.

Strassmann, J.E., Zhu, Y., Queller, D.C., 2000. Altruism and social cheating in the social amoeba *Dictyostelium discoideum*. Nature 408, 965–967.

Weinstock, G.M., Robinson, G.E., Gibbs, R.A., et al., 2006. Insights into social insects from the genome of the honeybee *Apis mellifera.* Nature 443, 931–949.

Chapter 76

2002 Social Networks

THE CONCEPT

A network of animals in interconnected by behavioral interactions. Each animal is a node within the network, and connections may be directly between pairs of animals, or indirect, via one or more intermediaries in the network. Networks help us to understand the flow of information in animal societies.

THE EXPLANATION

Network analysis provides extremely important tools for analyzing animal social systems. Network analysis has its roots in the desire of telegraph and telephone companies to build efficient communication systems which had the ability to handle surges of information and the redundancy necessary to cope with local network outages. National and international power transmission grids are built based on the same principles of capacity for handling demand and protection from failure.

Biologists and social scientists started using pieces of network and information theory in the 1960s based on principles developed decades earlier in the telecommunications industry (see Chapter 19: 1948 Information Theory). However, the major burst of incorporation of network theory into the analysis of social and community processes did not occur until much later. The 2002 Girvan and Newman paper is highly cited and serves as a landmark in adding network theory to the arsenal of strategies used for analyzing animal social behavior. They were far from alone, though, in the application of these theories.

Girvan and Newman (2002) particularly pointed to the formation of communities with high degrees of connectivity within larger networks that are more loosely connected. For animal social behavior this is highly relevant to structures like herds, packs, schools, or troops, within which connectivity is very high. In many cases extremely highly connected family groups are nested within these social settings, so a hierarchy of connectivity results. Girvan and Newman (2002) suggested a technique for detecting communities

Conceptual Breakthroughs in Ethology and Animal Behavior.
DOI: http://dx.doi.org/10.1016/B978-0-12-809265-1.00076-9

within larger networks that has proven useful in identifying social groups in animals.

Another publication that has been highly influential in the use of network analysis in animal behavior is Wey et al. (2008): they concisely bring together an overview of how network metrics might be applied in analyzing animal social structures and propose many interesting ideas about the application of network theory. A recent outgrowth of network theory in animal social systems is the concept of the keystone individual (Modlmeier et al., 2014; see Chapter 80: 2014 Keystone Individuals).

IMPACT 6

This is a great idea that has been more difficult to apply than its originators might have conceived. While network analysis is important in studies of social behavior and published examples exist of the application of network theory to empirical data, there has not been the explosion in the concept since the Wey et al. (2008) paper that might have been anticipated. The keystone individual concept (see Chapter 80: 2014 Keystone Individuals) may give this approach new energy.

SEE ALSO

Chapter 19, 1948 Information Theory; Chapter 77, 2004 Behavioral Syndromes—Personality in Animals; Chapter 80, 2014 Keystone Individuals.

REFERENCES AND SUGGESTED READING

Girvan, M., Newman, M.E.J., 2002. Community structure in social and biological networks. Proc. Natl. Acad. Sci. USA 99, 7821–7826.

Krause, J., Croft, D.P., James, R., 2007. Social network theory in the behavioural sciences: potential applications. Behav. Ecol. Sociobiol. 62, 15–27.

Modlmeier, A.P., Keiser, C.N., Watters, J.V., Sih, A., Pruitt, J.N., 2014. The keystone individual concept: an ecological and evolutionary overview. Anim. Behav. 89, 53–62.

Sih, A., Hanser, S.F., McHugh, K.A., 2009. Social network theory: new insights and issues for behavioral ecologists. Behav. Ecol. Sociobiol. 63, 975–988.

Wey, T., Blumstein, D.T., Shen, W., Jordán, F., 2008. Social network analysis of animal behaviour: a promising tool for the study of sociality. Anim. Behav. 75, 333–344.

Chapter 77

2004 Behavioral Syndromes—Personality in Animals

THE CONCEPT

A behavioral syndrome is a predictable pattern of responses, such as being shy or bold, that an animal displays through its lifetime.

THE EXPLANATION

In 2004 Sih and his colleagues Sih et al. (2004) started a revolution in the interpretation of animal behavior by bringing the concept of behavioral syndromes—animal personality—to the foreground in animal behavior research. It seems pretty obvious that animals within a species vary in their behavioral approaches to the world. Anyone who has interacted with domesticated animals such as dogs, cats, horses, or cattle knows that not all individuals are alike in their behavioral inclinations. Experience with breeding such animals also shows clearly that there are familial components to the behavioral variation; parents and offspring are more alike behaviorally than randomly chosen animals. Until the 1990s, large segments of the world of scientists studying animal behavior seemed to ignore this common knowledge. Sih et al. (2004) crystallized the shift toward crediting animals with variation in personality.

This shift toward considering variation among animals in natural populations as important began in the 1990s. Unfortunately, a semantic and philosophical barrier to considering animals as having emotion-like motivations for their behavior ran deep in the community of scientists studying behavior. The concept of personality had been viewed as uniquely human, partly because some of the dimensions of personality involve emotion.

The training of scientists had included urging them to not to anthropomorphize their study subjects, and specifically not to attempt to project subjective emotions or feelings on animals. Scientists who did so operated at peril of being accused of "Disney-ifying" their interpretations. In part, Morgan's canon (see Chapter 11: 1894 Morgan's Canon) was to blame for

Conceptual Breakthroughs in Ethology and Animal Behavior.
DOI: http://dx.doi.org/10.1016/B978-0-12-809265-1.00077-0

the blindness of this approach. These factors had made it difficult to assign meaning to the variability in behavior.

Lima and Dill's 1990 paper on fear in animals (see Chapter 63: 1990 Fear) gave scientists room to think about fearful animals and the implications of fearful behavior in population biology. Neuroscientists were finding that the framework of neurotransmitters regulating emotions and mood in humans is an evolutionarily conserved set of mechanisms that operates similarly across the vertebrates (see Chapter 27: 1964 Dopamine and Reward Reinforcement; Chapter 51: 1978 Animal Models for Depression). Some leading scientists, such as Jane Goodall, argued forcibly that it was fine to assign names to study subjects and to regard them as having emotional lives.

All of this set the stage for Sih and his colleages (2004) to make the argument that animals display behavioral syndromes, analogous to what we consider in humans as personality. Sih et al.'s (2004) paper essentially gave permission to animal behaviorists to think about variation and behavioral responses without being accused of anthropomorphizing. Much of the subsequent attention in this area has been given to behavior along a shy–bold continuum, with studies showing that being shy or bold each have balancing advantages and disadvantages, so these contrasting behavioral syndromes can co-exist in a population.

IMPACT: 10

This major advance in understanding why animals vary in behavioral phenotype begs for integration with the neuroscience of anxiety, reward, and emotion. The outcome has the potential for being the most important research thread in animal behavior to emerge in the first part of the 21st century.

SEE ALSO

Chapter 55, 1980 The Risk Paradigm; Chapter 71, 1998 Self-Organization of Social Systems; Chapter 79, 2004 Public and Private Information; Chapter 76, 2002 Social Networks; Chapter 80, 2014 Keystone Individuals.

REFERENCES AND SUGGESTED READING

Bell, A.M., 2005. Behavioural differences between individuals and two populations of stickleback (*Gasterosteus aculeatus*). J. Evol. Biol. 18, 464–473.

Bell, A.M., 2007. Future directions in behavioural syndromes research. Proc. Royal Soc. B-Biol. Sci. 274, 755–761.

Bell, A.M., Sih, A., 2007. Exposure to predation generates personality in threespined sticklebacks (*Gasterosteus aculeatus*). Ecol. Lett. 10, 828–834.

Brown, J.S., Laundre, J.W., Garung, M., 1999. The ecology of fear: optimal foraging, game theory, and trophic interactions. J. Mamal. 80, 385–399.

Dingemanse, N.J., Wright, J., Kazem, A.N.J., Thomas, D.K., Hickling, R., Dawnay, N., 2007. Behavioural syndromes differ predictably between 12 populations of three-spined stickleback. J. Anim. Ecol. 76, 1128–1138.

Dingemanse, N.J., Kazem, A.J.N., Reale, D., Wright, J., 2010. Behavioural reaction norms: animal personality meets individual plasticity. Trends Ecol. Evol. 25, 81–89.

Koolhaas, J.M., de Boer, S.F., Coppens, C.M., Buwalda, B., 2010. Neuroendocrinology of coping styles: towards understanding the biology of individual variation. Front. Neuroendocrinol. 31, 307–321.

Lima, S.L., Dill, L.M., 1990. Behavioral decisions made under the risk of predation: a review and prospectus. Canadian J. Zool. 68, 619–640.

Reale, D., Reader, S.M., Sol, D., McDougall, P.T., Dingemanse, N.J., 2007. Integrating animal temperament within ecology and evolution. Biol. Rev. 82, 291–318.

Sih, A., Bell, A.M., Johnson, J.C., Ziemba, R.E., 2004. Behavioral syndromes: An integrative overview. Quart. Rev. Biol. 79, 241–277.

Smith, B.R., Blumstein, D.T., 2008. Fitness consequences of personality: a meta-analysis. Behav. Ecol. 19, 448–455.

Stamps, J.A., 2007. Growth-mortality tradeoffs and "personality traits" in animals. Ecol. Lett. 10, 355–363.

Stamps, J., Groothuis, T.G.G., 2010. The development of animal personality: relevance, concepts and perspectives. Biol. Rev. 85, 301–325.

Wolf, M., van Doorn, G.S., Leimar, O., Weissing, F.J., 2007. Life-history trade-offs favour the evolution of animal personalities. Nature 447, 581–584.

Chapter 78

2004 Maternal Epigenetics

THE CONCEPT

In addition to normal genetic effects, handed down via genetic material from parent to offspring, mothers may influence their offspring by affecting gene regulation. Maternal hormones, stress, and nutrition can all result in epigenetic effects.

THE EXPLANATION

It is difficult to put an exact date on when biologists realized that a mother's influence on her offspring might extend to changing how the young animal's genes are expressed. The paper I've chosen as the fulcrum, Weaver et al. (2004), was a key step in the proof of this concept.

In order to understand why epigenetic programing of offspring is such an important leap, we first need to understand what epigenetics are. Until around 2000, we understood that an organism has many genes, and that these genes turn on and off as needed. Part of the on/off programing of genes is developmental; some genes are active early in an animal's life and then are silenced, never springing back into activity. Promoter and repressor genes are partly responsible for regulating gene programing.

Then, scientists came to realize that changes on top of the genes (this is where the term epigenetics comes from) have a big role in long-term gene regulation. Primary among epigenetic changes are histones, proteins that wind around DNA, and methylation, the addition of methyl groups to DNA.[1] Genetic imprinting, silencing of genes by parental influences, is a different epigenetic mechanism and this seems to be limited to the placental

1. Epigenesis is why stem cells are so important in research on growing replacement tissues for transplants. The DNA in pluripotent stem cells, in particular, is relatively little modified by histones or methylation, meaning that stem cell genes can be more easily activated. Pluripotent stem cells have the potential to develop into many different tissue types; cells collected from already-determined tissues are epigenetically programmed to produce only cells of that tissue type and are difficult to reprogram.

Conceptual Breakthroughs in Ethology and Animal Behavior.
DOI: http://dx.doi.org/10.1016/B978-0-12-809265-1.00078-2

mammals. While epigenetic changes may be reversible, they often have lifetime effects on gene regulation.

In the Weaver et al. (2004) paper, mother mice were found to affect the epigenetic profile of their offspring via licking behavior and a nursing posture called "arch-backed." Mothers that licked their pups more and arched their backs when nursing affected histones and methylation in the pups' genome. These effects appeared after the pups were born, but many subsequent studies show that maternal epigenetic effects can occur prenatally as well.

Much of the research on prenatal epigenetic effects on offspring behavior has focused on the effects of stress and maternal behavior on pups in rats and mice (e.g., Braithwaite et al., 2015; Keverne et al., 2015), but there is every reason to believe that these models reflect general truths for vertebrates, and that prenatal epigenetic effects should be possible in most animals.

IMPACT: 9

That a mother's influence on her offspring extends to changes in the offsprings' regulatory genome before laying or birth has major importance in understanding behavioral phenotypes. The hormonal, nutritional, and social environments that affect the mother extend, via epigenetic influences, to impacts on the young. This is an exciting and very active area of research; the 2004 paper represents a starting point for an extremely important area of inquiry.

SEE ALSO

Chapter 44, 1974 Parent–Offspring Conflict.

REFERENCES AND SUGGESTED READING

Braithwaite, E.C., Kundakovic, M., Ramchandani, P.G., Murphy, S.E., Champagne, F.A., 2015. Maternal prenatal depressive symptoms predict infant NR3C1 1F and BDNF IV DNA methylation. Epigenetics 10, 408–417.

Cameron, N.M., Champagne, F.A., Fish, C.P.E.W., Ozaki-Kuroda, K., Meaney, M.J., 2005. The programming of individual differences in defensive responses and reproductive strategies in the rat through variations in maternal care. Neurosci. Biobehav. Rev. 29, 843–865.

Keverne, E.B., Pfaff, D.W., Tabansky, I., 2015. Epigenetic changes in the developing brain: effects on behavior. Proc. Natl. Acad. Sci. USA 112, 6789–6795.

McCarthy, M.M., Auger, A.P., Bale, T.L., et al., 2009. The epigenetics of sex differences in the brain. J. Neurosci. 29, 12815–12823.

Weaver, I.C.G., Cervoni, N., Champagne, F.A., et al., 2004. Epigenetic programming by maternal behavior. Nat. Neurosci. 7, 847–854.

Zhang, T.-Y., Meaney, M.J., 2010. Epigenetics and the environmental regulation of the genome and its function. Ann. Rev. Psychol. 61, 439–466.

Chapter 79

2004 Public and Private Information

THE CONCEPT

Public information is available for other animals, ranging from offspring to competitors to predators, to assess. Private information is concealed, and only revealed to select animals. The evolutionary tension is that animals may seek to uncover private information, making it public, in order to exploit what they discover.

THE EXPLANATION

Can an animal keep some information to itself and release other information more widely? A territorial bird might visit its nest furtively, taking behavioral measures to cloak its movements when approaching its nest, yet it may at the same time loudly advertise the location of its territory. The nest location, with the very valuable contents of eggs or nestlings, is a closely guarded secret, private information. The territory is a publicly announced holding; the calls that signal the presence of the territorial bird are part of its strategy to keep interlopers away (Fig. 79.1).

Concealment can be critical to an animal's success. A foraging mammal may be sly about its movements, not wanting to share information about food with others of the same species. A predator typically shields its presence from its prospective prey. Hiding information is widespread and evolving to pry loose information and exploit it is equally common.

Danchin et al. (2004) synthesized the theory behind public and private information and created a framework for investigating how information is kept private and how public information is exploited. A useful framework for categorizing acquired (as opposed to genetic) information is:

- private information that has remained private;
- private information that has been intercepted by unintended receivers;
- public information that is not concealed because it is low value; the cost of concealment would exceed the cost of release; and
- public information that is purposefully released with either honest or dishonest intentions.

Conceptual Breakthroughs in Ethology and Animal Behavior.
DOI: http://dx.doi.org/10.1016/B978-0-12-809265-1.00079-4

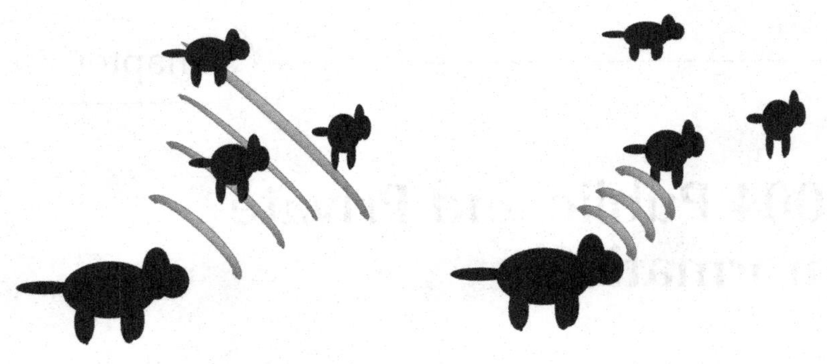

FIGURE 79.1 Public information is broadcast and can be received and assessed by any animal within the range of the signal. Private information, such as a whisper, is directed to one or a limited group of animals and hidden from the general population. Eavesdropping and signal interception are means by which animals may attempt to take advantage of information that is intended to be private.

All of these information types hold interest, but Danchin et al. (2004) drew particular focus to mechanisms of interception of information that was intended to be private. Eavesdropping and the exploitation of others' experiences by copying their habitat or mate choices became important research foci as the result of this paper.

The second part of Danchin et al.'s (2004) argument deals with cultural changes in public information. This aspect considers animals that live in groups, have overlapping generations, and the ability to shape their behavior by observing the behaviors of other members of the group. Among mammals, many primates, some carnivores, and perhaps some ungulates fit this model. Social birds, like sociable weavers, may also fit, but most examples of cultural evolution of public information in birds relate to learning songs. The cultural dimensions of private and public information have gained less attention. A very important avenue of future investigation will consider the interaction of cultural effects on the value of either keeping information private or making it public.

IMPACT: 7

Danchin et al.'s 2004 paper was an eye-opener for many animal behaviorists. It set the ground for explicitly considering how evolution shapes the control of information. Investigations of strategies for hiding information and counterstrategies for obtaining information became a major new research thrust. This topic also plays into human penchants for keeping secrets, sharing secrets and for sometimes violating boundaries by revealing other individual's private information.

SEE ALSO

Chapter 50, 1977 The Evolution of Mating Systems; Chapter 66, 1991 Receiver Psychology.

REFERENCES AND SUGGESTED READING

Clobert, J., Le Galliard, J.-F., Cote, J., Meylan, S., Massot, M., 2009. Informed dispersal, heterogeneity in animal dispersal syndromes and the dynamics of spatially structured populations. Ecol. Lett. 12, 197–209.

Dall, S.R.X., Giraldeau, L.A., Olsson, O., McNamara, J.M., Stephens, D.W., 2005. Information and its use by animals in evolutionary ecology. Trends Ecol. Evol. 20, 187–193.

Danchin, E., Giraldeau, L.A., Valone, T.J., Wagner, R.H., 2004. Public information: from nosy neighbors to cultural evolution. Science 305, 487–491.

Seppanen, J.-T., Forsman, J.T., Monkkonen, M., Thomson, R.L., 2007. Social information use is a process across time, space, and ecology, reaching heterospecifics. Ecology 88, 1622–1633.

Valone, T.J., 2007. From eavesdropping on performance to copying the behavior of others: a review of public information use. Behav. Ecol. Sociobiol. 62, 1–14.

Chapter 80

2014 Keystone Individuals

THE CONCEPT

An animal that stands out in a social group as an information center, leader, or dominant is a keystone individual.

THE EXPLANATION

A keystone individual is central to its social network. It has a disproportionate effect on the animals in the group. This concept, first fully developed in a review by Modlmeier et al. (2014) is constructed around the various keystone roles animals can play in a society. These roles differ in how they develop, how long they last within the group, and their impact on the behavior of other animals, yet they are unified by the importance of their influence on the group. By exerting a large influence on the behavior of the group, the keystone animal shapes the actions and personality of its network.

Modlmeier et al. (2014) gathered characterizations of keystone individuals such as: broker, catalyst, elite, hyper-aggressive, leader, tutor, and so on. This list partly captures the spirit of the concept, but the deeper point is that social information and the direction of group-level choices often revolves around a single individual. Sometimes the identity of the keystone animal is predictable—a dominant animal or an experienced animal, for example. In other cases, the keystone role falls to an animal that is in the best location to gather the information needed. Keystone individuals within societies show up in many different species, ranging from primates to insects.

One differentiation made among types of keystone individuals is activator versus information center. An activator instigates change in the group, such as moving from inactivity to predator evasion. An animal in an information center role may have knowledge based on experience or may gather information from others in the group as it comes in. A dancing honeybee exemplifies an information center; she has gathered information from the environment and from other bees about a food location and then she dispenses that information to other bees that are waiting to forage.

Conceptual Breakthroughs in Ethology and Animal Behavior.
DOI: http://dx.doi.org/10.1016/B978-0-12-809265-1.00080-0

Focusing on keystone animals is one a way of applying network analysis to understanding information flow in social groups (see Chapter 76: 2002 Social Networks). When considering hypotheses about keystone animals, investigators generally also study the rest of the network, as there is value in looking at the routes by which information travels in groups, which can include direct communication as well as transmission via one or more linking animals, and at animals that serve as secondary nodes within the network.

IMPACT: 4

The exact impact of such recently published work cannot be known, but the integration of network theory, behavioral syndromes, and the keystone individual concept is highly likely to be a highly productive route forward in understanding animal social systems. In future work, the keystone individual concept will make scientists more aware of the potential for animals to play roles as information centers, activators of group behavior, and group leaders.

SEE ALSO

Chapter 76, 2002 Social Networks; Chapter 77, 2004 Behavioral Syndromes—Personality in Animals.

REFERENCE

Modlmeier, A.P., Keiser, C.N., Watters, J.V., Sih, A., Pruitt, J.N., 2014. The keystone individual concept: an ecological and evolutionary overview. Anim. Behav. 89, 53–62.

Index

Note: Page numbers followed by "*f*" refer to figures.

Z

Printed in the United States
By Bookmasters

Printed in the United States
By Bookmasters